REALSPACE

REALSPACE

*The fate of physical presence in the
digital age, on and off planet*

PAUL LEVINSON

Routledge
Taylor & Francis Group

LONDON AND NEW YORK

First published 2003 by
Routledge

Published 2013 by Routledge

2 Park Square, Milton Park, Abingdon, Oxfordshire OX14 4RN
711 Third Avenue, New York, NY 10017

First issued in paperback 2016

Routledge is an imprint of the Taylor & Francis Group, an informa business

Typeset in Weiss by M Rules

British Library Cataloguing in Publication Data
A catalogue record for this book is available from
the British Library

Library of Congress Cataloging in Publication Data
Library of Congress Cataloging-in-Publication Data
Levinson, Paul.
Real space: the fate of physical presence in the digital age,
on and off planet / Paul Levinson.
p. cm.
Includes bibliographical references.
1. Astronautics and civilization. 2. Outer space–Exploration–
Social aspects. 3. Cyberspace. 4. Information technology.
5. Computers and civilization. 6. Internet. I. Title.

TL794.5.L48 2003
303.48'3–dc21
2002037154

ISBN 13: 978-0-415-75377-7 (pbk)
ISBN 13: 978-0-415-27743-3 (hbk)

To Columbia and her crew, 2003

CONTENTS

CONTENTS

CONTENTS

PREFACE
Four departures

Realspace is the third book about communication and technology I have published with Routledge in the past six years. But it differs from the first two – *The Soft Edge: A Natural History and Future of the Information Revolution* (1997) and *Digital McLuhan: A Guide to the Information Millennium* (1999) – in four crucial ways.

First, this book is a critique of the digital age – at least, in its current state. Whereas the previous two volumes were designed to explicate the digital revolution – call attention to its importance, trace its origins in human history, logically sketch its likely future – *Realspace* examines what needs to be corrected. It looks at what's wrong with life online. Or, to be more precise, it looks at fundamental aspects of human life that cannot be satisfied by working and playing and shopping in cyberspace.

This might seem like a reversal of my previous enthusiasm

about digital communication, but actually it is not. My "anthro-potropic" theory of media evolution (anthropo = human; tropic = towards) – first developed back in 1979 in my "Human Replay: A Theory of the Evolution of Media" – suggested well before the digital age that as media evolve, they become more human, less artificial in their performance. Thus, photography and television evolved from black-and-white to color because we see the world, prior to and without media, in color. There is much to commend in the Internet from this perspective. E-mail, to cite one example, humanizes traditional letter writing by permitting the written word the kind of immediate delivery we have in the natural, unmediated world with face-to-face speech.

But if we came to rely on e-mail to the exclusion of speech, this presumably would not be a good thing. Not because e-mail might replace phone calls; in that case, one medium would only be ousting another. But were e-mail ever to totally replace in-person conversation, the result would be a net loss, not increase, in natural communication. For however more human the flow of e-mail than traditional paper mail, the new, digital flow would be significantly less human than face-to-face conversation.

Cyberspace, in other words, exceeds its humanity when it challenges the core of realspace, or the place where physical presence, not just exchange of information, is essential. The thrust of this book is that although complete supplantation of realspace by cyberspace will never happen – precisely because touching, feeling, and moving through the real world are too intrinsic to our lives – we nonetheless need to be aware of the damage that can result if the scale tips too far towards cyber-space, or the transport of fleshless information instead of people.

But *Realspace* is not only a warning about the excesses of cyber-space on Earth. It is an advocacy of what we should do in the realspace beyond our planet – a call to action in outerspace.

Neither of my previous two books were works of advocacy. *Digital McLuhan*, to be sure, advocated the importance of Marshall McLuhan, but it was more a defense of his work and an assessment of why it is so necessary in these times. *The Soft Edge* had some chapters of advocacy – in particular, for freedom of expression on the Internet and for copyright in the digital age – but it was mostly history, analysis, and projection.

On our planet and in our everyday lives, realspace is here for the taking, always and obviously at hand, and in that sense needs no special pleading. Encroachment upon it by cyberspace will eventually be self-correcting; the only question is how long and with what ill-effect. But space beyond our planet is something else. Though it is also out there for the taking – for our explo-ration, and eventual settlement if we choose – we have to make an enormous investment in technology to get there. Absent such a commitment we almost certainly will not, in any lasting way.

Realspace thus makes the case for outerspace. What are the mainsprings of our progress off of this planet thus far, and why have they failed to take us further? Why is living in any place other than on Earth so crucial to our self-understanding and survival? How can space be more attractively packaged? What wellsprings of support for solar civilization are untapped?

Information technologies – not only cyberspace, but tele-scopes and robots in many of their applications – enter this campaign again as two-edged swords which can clear the path as first scouts, but are no substitutes for human beings in the long run beyond this planet.

Realspace is a departure for me in a third way. Although the argu-ments I offer are, I hope, informed by decades of thought and research, the book is not, I would say, a traditional work of scholarship. Rather, its purpose is to alert and convince as well as inform. And although there is nothing incompatible in these goals – the most persuasive arguments are usually well informed – the emphasis on persuasion rather than education means that some of the typical accoutrements of academic scholarship will be limited in this book.

Thus, although sources of information will be identified in the text, I will forgo the usual scholarly practice of citing multiple sources for a single topic, and in general will not provide exten-sive citations of arguments similar and opposed to those in the book. The work is not intended as a comprehensive overview of research in the field.

Instead, the reader will find a small, select Bibliography at the end of the book, fully annotated. These sources are diverse, ranging from Plato and Freud to Rokusuke Ei and Hachidai Nakamara, writers of the misnamed "Sukiyaki" hit record of 1961. Given that one of the central themes of this book is out-erspace as the next step in the evolution of Earth and all of our culture, such diversity of sources is not surprising.

Finally, *Realspace* is different from any book I have written before for a reason I could not have known when I began the book. It is the first book I finished writing after September 11, 2001.

I am a New Yorker, and the attacks hit home in more ways than one. In terms of this book, I think the terrorism of the twenty-first century has given a keener edge to the need for human civilization beyond Earth as soon as possible. On the one hand, space of course will always be there. On the other hand, the capacity of any society on Earth to get there remains precarious, because it is vulnerable not only to attack but essential response to attackers and would-be attackers, which can divert resources intended or available for space. Meanwhile, the advent of anthrax in the mail puts the issue of e-mail versus paper mail in yet another light.

These and related concerns are noted at various places in this book, which I conclude with a special chapter, "Realspace in an Age of Terrorism" (Chapter 11).

The arguments for fuller human presence and interaction, both on and beyond this planet, were strong to begin with, and are even stronger today.

NEW YORK CITY
FEBRUARY 2003

1

BICYCLING INTO OUTER SPACE

Realspace is about the need for full, face-to-face interaction with people and the world around us in the age of the Internet. It is about life before and beyond cyberspace. It is therefore also about the need for human involvement in outer space – the space not only on our planet but beyond it.

We enter the world talking and walking. Most of us are alive nearly a year or more before this really occurs, but sooner or later each of us proclaims our humanity by toddling up to someone and saying "Mommy" or "Daddy" or the equivalent in languages other than English – by walking up to someone and talking. Those of us able to talk but not walk, or walk but not talk, require our special concern, but fortunately they are few.

This marriage of communication and transportation in humanity continues in our technology. In the first three-quarters of the nineteenth century, telegraph poles and wires sprouted up alongside railway lines. It was easier to seek fortunes far away when feelings could be conveyed home at the speed of electricity. At the end of that century, telephones and bicycles began allowing us to speak and move through the world more fluidly, less formally, without telegram deliveries by third parties, and the set schedules of trains. We finally broke entirely free of the umbilical cords of both rails and wires in the twentieth century, with radios that spoke and sang to us everywhere, and cars that we could drive just about anywhere anytime we pleased. Television and airplanes later helped expand this mobile life to global proportions.

But as we embark upon the twenty-first century, the historic partnership of communication and transport has become unhinged. We talk on telephones that we carry in our pockets. The world's libraries are within reach of our fingertips on the Internet. Information, it seems, is available everywhere at the drop of a hat. Where is our equivalent leap in transportation?

In the 1960s, outer space looked likely to be our next destination. But since we walked upon the Moon, only a tiny fraction of the world's population has made it off of this planet. In contrast, millions work and play every day on the Web. Unlike past revolutions in transport which resulted in just about everyone driving and millions flying, the Space Age has so far been peopled by lonely travelers braving the dark.

Realspace is, in part, about how this discrepancy arose. It is about how the ancient union of talking and walking, movement of our information and movement of our bodies, became unglued. It is about the keen biological need that we have for transport through actual space, not cyberspace, and how we can accommodate this requirement of our humanity. It is about the danger that awaits us if this need remains unrequited.

The limits of cyberspace

How much of the world do you get when you walk along the shore at sunset? If you walk barefoot, you feel various textures of sand beneath your feet. Even if not, you see colors in the sky and sea, blues and greens and purples and reds and everything in between. Cries of seagulls crease the air. You feel the breeze in your hair, on your face, on your body, with the warmth of the departing sun as accompaniment. You smell the sea. You can even taste it.

How much of this is available to the stroller in cyberspace?

Most of the colors, of course. And let's grant many of the sounds. And your screen may even display colors that do not quite exist along real shorelines, but are nonetheless pleasing and even appropriate to a coastline view.

And then there's the greatest advantage of all for the traveler online – it can bring you to any number of shorelines, regardless of where you and they may be. A virtual infinity of shorelines, equally and easily in view, from desert and mountain, from office and apartment. All you need is a computer screen, connected by wire or cell, to some appropriate faraway computer and camera.

That's impressive indeed.

Yet none of the online shorelines have breezes. None have setting suns that coat your body with receding warmth, charge the air around you with imminent evening.

How important is a breeze in your life? A sun in full-dimensional, multi-sensory, slowly setting splendor? Not very important when life is looked at in one way. But also all-important when life is looked at in another.

And what about the myriad flavors and other sensations that come to us through taste and smell and touch? There is no doubt regarding those – they are flat-out essential to our lives.

And the Internet provides not one of them.

Not that the Internet prevents us from enjoying the beach, or any of the tastes and textures of the world. Far from it. Access to the world's information, whenever and wherever we want it, can in fact give us freedom to enjoy the sand and the breeze and the water and whatever physical pursuits may please us. My family and I have borne witness to this just about every year since 1984, when we discovered we could stay much longer on Cape Cod if we brought along a computer that kept us connected to our business in New York and the world.

But the Internet frustrates as it liberates. Humans cannot live by information alone – in fact, the more we have of it, the more we want to touch, hold, taste, walk upon, actually move through what the information shows us. In the end, imagery and scenery cannot suffice. Representations are not enough. Life is a full-contact sport, and we crave our payoffs in actual reality.

Cellphones improve upon the Internet in this regard, by satisfying our need to be away from our desks and screens and out in the world. Their rings may annoy others, and distract us briefly from the world at hand, but their popularity continues to increase because they allow us, in good times, to have the cake of our information and eat it in the real world of transport, trees, and sidewalks. And in bad times – in an age of terror – cellphones can perform heroic, miraculous services. They allowed passengers on the hijacked United Airlines Flight #93 on September 11 to speak to their loved ones one last time, and to prevent that plane from being used for the horrendous ends of the three other hijacked planes on that day. Whether vehicles of convenience or life-savers, cellphones and their connectivity in the world at large are the first high-tech acknowledgment of realspace in the age of cyberspace. Where the choice was once communication, indoors, away from the physical world, or movement and transportation out in the world with no communication, cellphones open up a third possibility – the world outdoors with full communication.

The "Segway" – the pogo-stick scooter device, powered by battery and steered by muscle attitude – is another expression of our need to physically navigate the world in an age of cyberspace. Unveiled with fanfares at the end of 2001, the Segway promises to get us much closer to street life than even motorcycles or convertibles. Whether it keeps those promises or not, the Segway's introduction five years after the Web burst into public notice in 1996 is no coincidence: people wrapped in innumerable threads of gossamer information want out – out of the cocoon, out in the world. And they want out in a way that gets as much of the breeze in their hair, the sun on their face, as possible. The more information we get, the more we need of physical reality.

Realspace is about these transactions in the outside world –

often commonplace, sometimes extraordinary, always profound – that the Internet cannot make possible. Although real
space of course existed before the Internet, realspace – and the
need to look at physical transactions as perhaps an endangered
species – is a consequence of the Web. We might say that cyberspace brought realspace into focus. The world of cyberspace
requires realspace.

This book is about the importance of these real experiences
here on Earth, and in the realm of space beyond our planet.

The lure of outer space

If immersion in the physical realities of our world is so good,
who needs outer space? Why not just revel in the hearth of
Earth when we're not online?

Stand on the coast of California and look out over the ocean
and you'll know why. At the end of America's westward expansion, beyond the close of our frontier, is the past – the Old
World, the Eastern Hemisphere, on the other side of the Pacific
Ocean, where our species and civilization first arose. If you want
something genuinely new, the only way to go is up.

And what will we find there?

Increases in scientific knowledge, access to vast resources,
opportunities for adventure – all of these are important. Even
more crucial is what outer space does for the precariousness of
our planet, vulnerable to an asteroid or other devastating strike
at any time. The portrayal of this danger in melodramatic
movies makes it no less real. The destruction it could visit upon

our planet would be far worse than anything a terrorist attack, even a deliberately spawned plague, could likely mete out. But whether humanly or cosmically sponsored, devastations would encounter a significant buffer in a humanity dispersed across the universe, or even just off of this world. Physical movement of at least a part of our species beyond this planet is the best strategy for human survival of any global disaster, whether of natural or human origin.

But there is also a yearning for space that comes from a different place in the human psyche. It goes beyond knowledge of this or that planet or star, or even dodging our extinction in the wake of a hell-bent asteroid.

It comes from wanting to know who we are. From wanting to know more of just what we are doing here in this universe. Trying to answer that question from the sole vantage point of Planet Earth is like trying to understand any city just by peeking out of a window of an apartment you have lived in all your life. Actually, trying to understand the universe in that way is even more disadvantaged: it is like being stuck in an apartment where not only you but your parents and grandparents and every person before you has lived all their lives.

Given this need to know more about who we are, given the sense of frontier in space that has always fired our imaginations, why have we been so cautious, so moderate, in our journeys into space in the past few decades? The blame for that rests not in cyberspace, though the seduction of fleshless information moving at astonishing speed has not helped.

What went wrong in outer space?

It is amazing that we made it out into space at all. Lifting ourselves totally off and beyond this planet is – as far as we know – without precedent. In contrast to just plain air flight, which was old news to birds, insects, and other winged creatures for eons before we joined them in our flying machines, our movement into space is apparently something really new under the sun. Or, at least, under ours.

Perhaps such extraordinary leaps are always followed by decades – perhaps centuries, even millennia – of indifference and coasting. None of us, after all, was around when fire was harnessed, when writing was invented, when crops and livestock were first domesticated. Historical records for such signal events are not available. Perhaps lack of progress in *their* immediate aftermaths was lamented just as some of us now lament our lack of progress in space.

But here we are, nonetheless, heirs to walking on the Moon a little more than a decade after Sputnik, followed by three decades of humans not setting foot on a single extraterrestrial body. It's hard to avoid the question: What went wrong?

Numerous culprits suggest themselves. JFK's assassination robbed the US space program of its most inspiring leader. The Vietnam War soured Americans on military involvement in anything, including space. Understandable concern for the environment spilled like caustic acid over technological endeavors of all kinds. The end of the Cold War removed one of the two prime movers of space exploration, the Soviet Union. This in turn eliminated one of the main motivations of the other prime mover, the U.S.A.

Whatever the specific reasons, they have a common denominator: continuing reliance on central government to finance, organize, and direct our space efforts. Nothing like that reliance figured in the success of railroads, telegraph, telephone, bicycles, automobiles, radio, air flight, television, the Internet, and cellphones – though, outside of the U.S.A., governments have played a more major role in the control of radio and TV (usually to the detriment of the mass media in those countries). Only in pursuit of outer space have people been so utterly dependent upon government, not only for initial funding and regulation, but for every subsequent step of the process. In the U.S.A., NASA has been the equivalent of the New York Central Railroad, Bell Telephone, the Ford Motor Company, AT&T, RCA, CBS, IBM, and Microsoft, all rolled into one.

This is a criticism not so much of NASA as it is of tying our future in space to the vagaries of even the strongest democratic government on Earth.

Dictatorships may sometimes work better than democracies in the staging of great developments. The Pyramids were built at the Pharaoh's behest. Alexander the Great conquered most of the universe he knew by sheer dint of his personal will and military command. Rocketry was first developed for practical use as a weapon of long-range war by Nazi Germany. And the totalitarian Soviet Union launched our Space Age with Sputnik. True, American democracy took the next huge stride by getting us to the Moon. But the general lack of accomplishment since then raises the worrisome possibility that the Apollo program may have been the aberration, not the norm, for democratic governments as spearheads into space.

9

Yet totalitarian regimes have been even less reliable than democracies in the long run of space exploration and habitation. Where are the Nazi rocket and Soviet space programs now? Mir, the Soviet space station – humanity's first long-term space habitat – fell from the sky in 2001, too broken, too unreliable, too clunky to be worth fixing. "Friendship," the new International Space Station operated by the more democratic Russia along with the U.S.A., is the hopeful, more enduring replacement.

What, then, is the alternative to government initiative in space? Are individual initiative and corporate seeking of profit better partners for fulfillment of our cosmic ambitions?

Billions are made and lost these days in e-trades of high-tech stock. But that technology is all about information – a product of no physical substance. It can pop like a balloon in an instant. It already has. The information economy will no doubt recover – and it will burst again.

The real estate of space, however, for all its current inaccessibility, has one big, real advantage: it is real. It consists of immensely large tangibles. *Realspace* will consider how such tangibles can be more attractively packaged.

But we will also examine another aspect of human life, a pursuit distinct from business, science, and military endeavors, which has motivated the movement of millions of people across continents here on Earth. If our deepest need to get beyond this planet is spiritual – to learn a little more about who we are in this universe – then we would do well to enlist religion as a springboard to space. We have seen, in the terror of September 11 and its aftermath, the awful destruction that religious belief can motivate. Can it help fuel the reverse – the lifting of our species off this planet?

Space repackaged

We live in a world in which web pages, TV shows, phone calls, newspaper headlines, and myriad information packets daily vie for our attention. In order for people to get excited about space, reports of our exploits off-world have to be entertaining, provocative, inspiring, extreme. They have to compete not only with entertainment, but with breaking news of terrorism and war. One might think that news from the literal edge of existence – news from the universe, news with perhaps some inkling about who and what we really are – would trump or at least be able to hold its own with any Earth-bound matter, however exciting or grave. But even though humans in space are always on the tip of danger, only their death commands our attention, and just for a short time. Perhaps the very immensity of the story in space, the awesome profundity of what it stands to reveal, makes it difficult to present in bite-size media segments – in split-second crawls at the bottom of cable newscasts, or longer ten-minute segments, or even hour-long programs, or on web pages with hypertext links that can ripple to infinity in principle, but in practice are only as good as the aggregate talent of the many people, usually unknown to one another, who have written and programmed the pages.

Yet these are the tools of the proselytizer of space. How can they be best employed?

Reportage of new worlds played a major supporting role in the Age of Discovery on Earth. Accounts of Columbus' voyage by his son Diego became bestsellers in cities across Europe in the 1490s, and no doubt helped make his voyage the one that opened the Western Hemisphere to European settlement. Leif Eriksson had already reached the New World – around 1000 AD – but accounts of his trip were whispered rather than written, recited rather than printed, and thus had no lasting impact on the progress of the world. Today we have ample media to communicate instantly, globally, permanently, fleetingly, colorfully about just about anything. The Internet – which I call "the medium of media" because it conveys radio, television, books, shorter texts, telephone conversations, e-mail – along with cable television has made the world a ready audience. So the need is not to find a new medium of communication, but a better way of communicating about space through media already in play.

We can begin with the very way we talk about our space vehicles. Starships are more enticing than space shuttles; cloud offices are more inviting than space stations.

A name such as "Mermaid Dune" – given to a Martian rock site identified in NASA's 1997 Pathfinder mission – is a minor step in the right direction. It humanizes an alien environment, in contrast to 98-BLG-35 and the alphanumeric soup of appellations often given to new meteorites, asteroids, and even possible planets discovered in other solar systems.

The planets in our own solar system are already well named. They harken from an age in which our cosmos was viewed more humanly – seen as more of an extension of our lives than the current cavernous universe that science says we inhabit.

Realspace calls for a recalling of romance and myth and wonder in our scientific vision – not as substitutes for facts, but as better vehicles of their presentation.

Robots and missed golden opportunities

However attractively we may repackage space, it is still, undeniably, far less hospitable than any place on Earth. Dangers may lurk in an unknown lake, but it sells itself anyway as a potential place to swim, fish, or boat. Our experience in the cool, sweet waters of other lakes draws us in.

We have no such real experiences in outer space. Movement out there thus requires confidence in our technology – that it will be up to the task of making the space beyond our planet amenable to human life, just as we have been doing for millennia in inhospitable deserts and ice-caps right here on Earth.

A safe, less extreme approach to space would be to send robots. Computerized equipment can certainly relay lots of information. As a first wave, as an advance scouting party, robots certainly have their uses. And they do much more for our species in space than having no emissaries there at all.

But robots have severe drawbacks even as scouts. They can be blind to what they are not programmed to see, especially emanating from directions they were not programmed to scan. Absent some extraordinary breakthrough in artificial intelligence, robots are mostly cameras and pencils on stilts. They can return to Earth with real samples, but are by and large a technology of human communication rather than transport.

Viewing the landscape of Mars through robot eyes gives us a second-hand interaction with that environment, a representation, not the real thing. It is a form of television and cyberspace, and thus something less than full reality.

History teems with technology that stood on the verge of real accomplishment but never got off the ground. The Chinese invented the printing press at least half a millennium before Gutenberg, but no outpouring of books and newspapers ensued. They invented rocketry, too. The principles of motion photography were demonstrated in ancient Alexandria. The Aztecs independently invented the wheel, but it moved only their toys.

We stand at such an edge of opportunity now. We can turn inward, and merely continue to expand the parts of our lives that we spend talking, reading, shopping online. We can confine our global mobilizations – whether against terrorism, or combating natural calamities such as disease and starvation – to just improving our home here on Earth.

Or we can take this new access to huge amounts of information and this increasing global-mindedness with us, as we walk in the sunshine and rain of the real world, and seek our future in worlds beyond the sky.

We explore in more detail what talking and walking mean to our species and this view to the stars in Chapter 2.

2

WALKING AND TALKING

The reason they rhyme

On June 15, 1963, Kyo Sakamoto's "Sukiyaki" replaced Leslie Gore's "It's My Party" as the number 1 hit record in the U.S.A. The record is still the biggest international hit by a Japanese singer. Sakamoto was killed in the JAL plane crash of August 1985 – he was 43 – a harrowing tragedy that took the lives of 519 other people. But there is also a quiet aspect of cultural tragedy in the story of "Sukiyaki" which highlights the inadequacies of language in a vacuum, or communication removed from reality.

The recording was first released in Japan in 1961 as "Ue O Muite Aruko," lyrics by Rokusuke Ei, music by Hachidai Nakamura. The words have nothing to do with mixtures of noodles, vegetables, and meat served in a pot. The title translates into English as "I Look Up When I Walk." The first line reads, "I look up when I walk so the tears won't fall."

This might well have been an anthem for the need to look up at space rather than down at the Earth for our future – the theme of this book. It was not. But "I Look Up When I Walk" was a poetic love song whose lyrical import was lost in the "Sukiyaki" title. Most Americans not only had no idea what the song was about – Sakamoto was singing in Japanese – they were pointed in a nonsensical direction.

The great advantage of communication – especially our abstract language – is that it enables us to consider worlds of reality without coming into physical contact with its risks. Our ancestors who were able to say to their friends and family "there is a starving lion over the hill" did better than their non-linguistic relatives who had to pull their kin up the hill to point out the lion. Transportation without adequate communication is dangerous. Indeed, transportation of any kind, with or without communication, is intrinsically more perilous than communication. Planes, trains, automobiles and space shuttles can crash and kill their passengers. Books about those modes of transport pose no such threat to their readers.

But communication without context, words unleavened by physical presence, can be harmful too, certainly misleading. They are the stuff of misunderstanding, and lies. They can lead to wrong and even nonexistent paths. To the extent that we are living in a world of representations rather than realities, of seeing and hearing rather than touching, we are prone to confusions far more profound than taking "I Look Up When I Walk" to be a song about Japanese cuisine.

The biological antiquity of the coupling

The news that bacteria communicate – about when they have attained numbers sufficient to release toxins, exchange genes, produce antibiotics to kill fungi and other bacteria – suggests an intertwining of walking and talking at the very root of life. Indeed, insofar as DNA is a set of instructions, life is as much an act of communication as of locomotion.

Since information is intangible, however, we are more likely to notice movement of bodies in the biological world than movement of data. Everyone knows that fish swim, birds fly, and some insects do both, as well as scamper. The modes of communication of these organisms are less obvious, but they direct, buffer, and otherwise mediate the movement of these organisms at every turn.

The evolutionary psychologist Donald T. Campbell pointed out more than a quarter century ago that seeing, hearing, smelling, tasting are vicarious forms of touching, which allow living organisms, including us, to know something of the world without coming into direct physical contact with it. The amoeba, which presumably lacks even the bacterial gift of the gab, knows only what it bumps into. Should that unseen, unheard, unperceived object be toxic, the amoeba dies with its first and only knowledge of the object. Seeing and hearing – modes of perception and vectors of communication – offer escape routes for higher organisms, which can identify a danger from afar and move in an opposite direction (or, if appropriate, not at all).

But we and the sighted pay a price for this advantage: what we see, hear, smell, even taste may not be an accurate representation of the world. The water we see in the glistening sand may be refracted light; food that tastes bad may be nonetheless nutritious. In contrast, although the amoeba may die with its first and only embrace, it suffers no optical illusions. It suffers no illusions at all, because it lives – and dies – with no perceptions, no preconceptions, no modes of sensation from which illusions may arise.

This potential inaccuracy of all perception and communication is offset by physical engagement – by locomotion to and through the source of perceptions. If we want to really know if that shimmering image in the sand can quench our thirst, we need to travel to it.

The evolution of language in human beings, which allows us to communicate about objects, events, experiences in their complete physical absence – indeed, to communicate about notions such as the concept of concept that have no physical existence at all – has given us enormous advantages over the rest of the living world. Far safer to learn about a raging fire without smelling the smoke or seeing the flames. But language has also magnified our potential for error.

Human roads

We walked out of Africa talking. Earlier hominids apparently neither walked nor talked as well as Homo sapiens. Evolution improved the results of our transport via communication, by allowing us to talk about where we were going, and this more effective locomotion offered better reality checks on our communication. Movement of people and information were two sides of the same rolling coin. Transportation and communication worked – walked, talked – hand in hand.

After talking, the next great accomplishment in human communication was writing – "the voice of an absent person," as Freud put it. However, paintings on cave walls and hieroglyphics on monuments could not travel very far. They extended the disembodied voice across time, not space. It was left to the alphabet, written on papyrus, parchment, and paper, to bring the voices of absent people to the world at large.

The alphabet was remarkable for another reason: it was a totally abstract system of some twenty odd letters that could represent anything, precisely because its symbols looked like nothing. (Hieroglyphics, and modern Chinese ideograms, in contrast, were rooted in a specific symbolic image for each thing or idea in the world.) Easy to learn, enormously effective, the abstract alphabet was the palm-top, cellular, on-the-go, slang, hip-hop, shorthand system of its time. Indeed, it was and is digital in the same sense as DNA, since all derive their power from their utter lack of resemblance to the worlds they describe and control.

Our alphabet was invented by the Phoenicians – unsurprisingly a maritime, mobile civilization, with ships that plied the Mediterranean, the coast of Africa, and at least as far away as Britain for trade. The Phoenicians needed a communications "lite" to keep track of their numerous commercial transactions. The ancient Hebrews used the alphabet to describe a God who was omnipresent, omnipotent, and invisible – visually contradictory characteristics that were basically unwritable with hieroglyphics. The ancient Greeks argued about the value of the alphabet and writing – see Socrates' attack on writing in the *Phaedrus* – but Plato, Aristotle, and others used it to develop the still burgeoning fields of philosophy, science, and law. Neither the Hebrews nor Greeks of the ancient city-states traveled very far. But Alexander, Aristotle's student, carved out an empire from Europe to India, and literacy in the world he left would be higher than at any time until the nineteenth century.

The Romans understood balance in most things, including transportation and communication. Socrates had decried the written word in part because you could not ask it a question (it would give the same unvarying answer – what was already on the document); he understood the need for physical presence, and the interactivity it afforded communication. The Romans understood the need for physical presence in enforcement of its laws, and the pronouncements of its emperors. They built roadways on which messages, as well as merchants and men under arms, could be easily conveyed.

Some of the roads and the messages survived the fall of Rome. The roads endured in physical reality; the messages were transcribed by the Roman Catholic Church, the Byzantines, and Islam.

The printing press ignited the Renaissance. It did for the alphabet, five and half centuries ago, what the alphabet had done for communication two or more millennia earlier. In both cases, lives and societies were drastically transformed. The Protestant Reformation, the rise of national states, the scien-

tific revolution, the advent of public education and of capital-
ism, all fed on the new ubiquity of writing that flowed from the
press. ("Just as one voice can be heard by a multitude of ears, so
single writings cover a thousand sheets," Stradanus captioned
his contemporary engraving of a bustling sixteenth-century
print shop.) Advances in transport were also made. But it
would not be until the nineteenth century that transportation
would catch up and then move step by step with inventions in
communication that by then would be happening on almost a
decadary basis.

Railroads

Improvements in transportation were made in the Early Modern
Age – the introduction of the wheel to steer ships in the seven-
teenth century was a major accomplishment – but they refined
rather than replaced the natural mainsprings of transport in use
since antiquity: horses and wheels on land, winds on sea. The
harnessing of Watt's steam-engine for rail and water travel in
the first part of the nineteenth century changed all of that. The
Industrial Revolution, which had come to communication with
the mass production of texts from the printing press, was now at
the service of transportation.

The steam locomotive had a more immediate impact than
the steamship, and perhaps more in the long run as well. As late
as the 1830s, the British Admiralty saw no naval future for
steam-powered vessels, and windsails remained the preferred
mode of movement for much of the century. In contrast, major

and minor cities were sewn together with rail lines in the first half of the nineteenth century. And with the completion of the first American transcontinental railroad in 1869, the whole of the U.S.A., coast to coast, was traversable in five days. The natural circulatory system of rivers was now augmented and exceeded by a technological grid that moved people and materials farther, faster, smoother, and more reliably.

Information was moving farther and faster by then, too. The completion of the transatlantic telegraph cable in 1866 surpassed the reach of any railway, linking hemispheres together across an ocean. The telegraph's reach lay in electricity, and its capacity to carry information at the speed of light. Just as talking had initiated humanity, and the printing press the age of mass production, so the telegraph inaugurated the mass harnessing of electrical energy – which in the twentieth century would light most of the world, as well as enable it to talk, see, and hear across vast distances. If the railroad gave nations an improved circulatory system and heart, the telegraph provided a better nervous system and brain.

But for all the distances – physical and informational – vanquished by the railroad and the telegraph, they also shared a profound limitation that went against the grain of free-wheeling, hunting-and-gathering humanity: both required fixed systems, of enormous cost, to work their wonders. If you lived far away from a railroad station and a telegraph office, if you were off-track and beyond the wire, you were out of reach and out of luck for most of the nineteenth century.

Out of the box

The telegraph was not the only communication marvel in the first part of the nineteenth century. Photography was invented at almost the same time – and, in the case of Samuel Morse, who worked on both, almost by the same inventor – and did for time and preservation of information what the telegraph did for distance and speed. (Morse came close, but was never able to perfect a photographic process – that is, a chemistry for images that would last. That accomplishment belonged to Daguerre.) In one sense the photograph was even more "set" in its ways, more stuck in place, than the telegraph – for the photograph took the world as it was and committed it to plate and paper. That, indeed, was precisely what the photograph was intended to do. So, unlike paintings and telegrams, whose contents were wholly at the discretion of their creators, the photograph was a faithful transcriber of the real. But once developed, a photographic image could be carried anywhere, anytime, by foot, horse, rail, or any other means of human conveyance. For all its fidelity to a specific, tiny piece of the world – to the object of its image – the photograph required neither rails nor wires to move anywhere its possessor pleased, after the image had been captured and fixed in time. As André Bazin observed in the 1950s, the photograph no doubt rescues an image from its "proper corruption" in time. But we can observe further that, once taken out of the flux, the image on the photograph is inserted back into a world in motion via the simple expedient of placing it in your suitcase, handbag, or pocket.

Eventually – with George Eastman's invention of the Kodak

camera in 1888 – the photograph could not only be carried but created anywhere. Visits to the professional photographer's studio were no longer necessary. In fact, the professional photographer was no longer necessary, except for ceremonial occasions; the whole world and its contents were now fit to be captured (or, at least, its images) by anyone with a little box camera.

Meanwhile, the telegraph was on its way to supplantation by a device which, while still requiring wires, dispensed with the need for telegraph offices, and recitation and delivery of messages by third parties. The invention of the telephone in 1876 brought the distant human voice directly into the home. This shattered the Victorian notion of the home as a safe haven from the business of the outside world, even as it made the home safer, since folks within could now call out for medical or police assistance. Either way, there was now less distinction, in terms of informational access to people, between inside and outside – the home and the world at large had much more in common. Moreover, the voice of the speaker went directly into the ear of the listener on the phone. This mouth-to-ear contact made the telephone the most virtually intimate medium in history (it continues in that position even today).

Modes of transport would soon follow suit, first in the bicycle. Introduced with pedals and handlebars in Scotland in 1839, the bicycle took a more modern turn with wheels of equal size in 1885, and pneumatic tires in 1888. In an utter departure from the mass formality of the train, the bicycle allowed people to come and go as they pleased, on their own schedules, person to person. Even before the bicycle built for two, this was a revolution in interpersonal transport. And the bicycle led to improvements in roadways.

Wheels were thus already in high gear in the early decades of the twentieth century, in the rotary dials of telephones and the spinning spokes of bicycles. But the world still yearned for long-distance communication without any wires at all, and for a vehicle of transportation that offered the flexibility of the bicycle and the power of the train.

Impulse power and partners

For all its charm, simplicity, and humanity, the bicycle was unsuited for long-distance transport, carrying of passengers, and even moving safely after dark or in the rain. Trains did that, as did horse-drawn vehicles, but automobiles did it better. Invented in Germany in the mid-1880s, brought to America and refined by Henry Ford at the turn of the century, the automobile soon excelled not only in train tasks but in shorter bicycle jaunts, house to house, house to business. The automobile – personal, powerful – soon became the preferred mode for just about every kind of short-to-medium-distance land travel. It injected people into the world, en masse yet individually, into every nook and cranny, like no other transport device before. And by the mid-twentieth century the car had become the moving locus for a revolution in communication: rock 'n' roll radio.

Marconi received his first transatlantic wireless signals in 1901, two years before Ford started his motor company. Marconi had invented the wireless – which communicated via electromagnetic carrier waves, which required no fixed conduits, indeed no conduits of any kind – in the teeth of pronouncements from Heinrich Hertz, discoverer of electromagnetic waves, who had advised that use of the waves for radio transmission would require a broadcast tower the size of Europe, and therefore was not possible. But Marconi's invention was nonetheless not to be the revolution in telegraph and telephone that he had envisioned. Broadcast equipment was in fact much more expensive than receivers, and by the 1920s it was clear that what Marconi had brought into being was not a tetherless telephone system but a one-way, instantaneous mass audience of simultaneous listeners. In the nineteenth century, the mass transit of rails had been complemented by the interpersonal communication of the telegraph and telephone. In the first half of the twentieth century, the interpersonal transport of the automobile was complemented by the mass communication of radio.

After the Second World War, the complementary relationship became a full-fledged marriage. Although telegraph and telephone wires and poles had accompanied tracks, and eventually telephone calls could be made from railway stations, none could be made – until the end of the twentieth century – from trains themselves. In contrast, radios were introduced in automobiles in the 1920s and were standard equipment by the 1950s; "Sukiyaki" was no doubt heard in as many cars as homes in 1963. Cellphones would make their first big splash in cars in subsequent decades. The automobile, which could move people almost anywhere, any time, had also become a center – a center in motion – of communication. Walking and talking – transport

26

and communication on dirt roads – had been extended to the sleek highway, with the ancient partnership still very much intact.

But the one area the automobile and its growing package of communications could not bridge were the seas.

Global villages

We last saw sea power when we noted that sails persisted a lot longer than horses as vehicles of transport in an age of steam. The sinking of the *Titanic* in 1912 taught the world a new lesson about the limits of ships propelled by engines. And, just for good measure, a lesson about the limits of wireless communication was thrown in as well: the *Titanic's* telegraph saved all too few of her passengers (because the nearby *Californian*, in rescue range, failed to pick up the *Titanic's* distress call). The fates of transport and communication were linked in death as in life.

Industrial power and electric communication would continue to move and guide ships at sea, but there was already something new in the air – literally. The *Kitty Hawk* had flown for forty-five seconds in 1903, the same year that Ford started making automobiles. Air travel progressed slowly from stunts to daredevil flying to mail delivery – it took a long time to get off the ground as a vehicle of everyday long-distance transportation. And just as steam had been underestimated as a source of military power at sea, so too was air power over land, even after the First World War. But the Second World War settled that issue decisively, and by the 1950s, airplanes were beginning to replace ships at sea and many of the rails on land as vehicles of long-range mass transport. The automobile (and its bigger cousin the bus) had already outstripped the rails for short-distance transportation in the U.S.A., except in a few bigger, older cities such as New York and Chicago, where subways and "els" (elevated train lines) continued to thrive. In the second half of the twentieth century, cars and planes became the increasingly preferred modes of travel in America. Similar inroads were made in Europe, though it never came close to abandoning its rails, probably because the shorter distances of its major cities reduced some of the advantages of air travel.

Meanwhile, something had happened to radio as well. By the mid-twentieth century it had sprouted pictures – it became TV. Many at the time thought this would be the end of radio; just as "talkies" had obliterated silent movies a little more than two decades earlier, why would anyone sit around and listen to radio (sound only) when television (sounds and images) was equally available? At first, radio indeed sputtered. Jack Benny and *Gunsmoke* and all manner of radio stars and hit series jumped ship to TV; advertising revenue for radio plummeted. But radio hung on, and in the game of musical chairs it found itself playing, radio turned defeat into spectacular success: millions watched television rather than listening to radio from their sofas, but they then got into their cars and rocked along with radio on the road. Hearing without seeing, unlike seeing without hearing, turned out to be a profoundly human mode of perception. The world grows dark every night but never really silent; our eyelids shut off our vision of the outside world when we sleep but we have no earlids; to look at one thing is usually not to look at anything else, but, regardless of what we're looking at, we're usually able to listen to anything we like. Radio capitalized on these realities of perception and found a handy new vehicle in rock 'n' roll.

Further, radio was far better suited than television to the automotive audience. Imagine trying to watch television when sitting not in an easy chair but in the driver's seat of your car – where you can look at the TV screen only by not looking at the road. You would probably not get as far as the first commercial. But with radio on board you can listen and drive forever.

This era is what Marshall McLuhan saw when he noted that the world was becoming a global village. It was a time of television in the center stage of every home, in the living room, perhaps in the bedroom, and radio everywhere else, in the kitchen, the bathroom, the office, the automobile, and (courtesy of the replacement of bulky vacuum tubes with lightweight, small transistors) even in the street. And, when people turned to ABC, CBS, or NBC in the U.S.A., they saw the exact same thing if they were watching the same network, whether they were in New York, Chicago, Los Angeles, or a city a fraction of their size. And when people listened to a "Top 40" radio station for any length of time – say, a thirty- or forty-minute drive across town – they were bound to hear mostly the same songs.

So, from a communications standpoint, this world was much like a village, in which everyone hears the same town crier. But it was not quite global, and not even entirely like a village either as yet, because the television audience was national not worldwide, and its members were not regularly able to communicate among themselves. Satellites would soon widen the circle, and the Internet supply the interactivity, of these incomplete global villages.

The beginnings of space travel and cyberspace

Rockets and computers both burst forth in the Second World War. The first was not quite enough to win the war for Germany. But the second, in the form of superior Allied information technology and its cracking of the German Enigma Code, was enough for Germany to lose.

Thus, in their infancy and childhood – from the Second World War through Sputnik and Apollo astronauts on the moon – space and cyberspace continued the linkage of walking and talking which had animated railroad and telegraph, bicycle and telephone, automobile and radio, airplane and television. These couples were far from perfect unions or partnerships. At times – indeed for much of the nineteenth century – communication outshone transportation, with the telegraph, photograph, telephone, and phonograph offering possibilities more radically new, more drastically different from preceding kinds of communication, than the railroad, steamship, and bicycle were in the realm of transportation. But powerful revolutions in moving people were not that far behind, and in the twentieth century the automobile and airplane probably pushed transport even a little ahead of communication, although the impact of radio and then television made it a close call. In any case, throughout most of the last two centuries, as in the entire history of humanity, transportation and communication were more or less neck and neck.

Although information technology – in the form of Alan Turing's code cracking – had a more decisive impact on the course of the Second World War than rocketry, in the 1950s and 1960s it looked as if rocketry, missiles, and space travel would gain the upper hand. In those days, Sputnik to Apollo were certainly accomplishing more – in actual accomplishments as well as grabbing headlines – than anything produced by mammoth-size computers. And, militarily, intercontinental ballistic missiles and their shorter range siblings certainly played a more crucial role – as in the Cuban Missile Crisis in 1962 – than the nascent Internet (born as the U.S. Army's Arpanet) at the end of that decade.

In Chapter 3, we will consider how the extraordinary success of the Internet in the past decade, and the surprising stalling of the space program in the aftermath of Apollo, overturned that trend and now threatens to unhinge the entire balance.

3

BREAKING OUT OF WINDOWS AND CYBERSPACE

Apollo 17's voyage to the Moon in 1972 would be the last time people would set foot on that natural satellite in the twentieth century. The cancellation of the Apollo program in the Nixon administration had made it clear to observers, even then, that humans would be a long time in getting to the Moon again.

Among the most interesting reactions to this sad moment was William Irvin Thompson's. The loss of physical access to the Moon for a few more years, Thompson argued in *Passages About Earth* (1973), was not all that important. What counted was the new perspective about Earth, life, and humanity that looking back on our world from space and the Moon had already given us. We already had the information, the revelation, Thompson suggested – so continued physical access did not matter. If we took a century or two to return to space via anti-gravity devices instead of chemical rockets, that would be OK.

When I first read Thompson, I did not fully realize that the pages presented an early plea for cyberspace over realspace, for the pleasures of information and representation, for the contemplation of things distant and absent, over direct interaction with reality. I do remember thinking how strongly I both agreed and disagreed with his analysis. The philosophical, cosmological, ecological, spiritual implications of space travel were as important to me as to Thompson. But how could these be nurtured and developed on the basis of memories and replays?

The success of the *Star Wars* trilogy in the movies, and the revived *Star Trek* on TV, seemed to support Thompson's view in the ensuing decade. Interest in outer space remained high, even though no Apollo or other astronauts ventured to and from the Moon.

But by the mid-1990s a new kind of audio-visual medium was rapidly taking hold. On the one hand, it crystallized and even epitomized the emergence of the global consciousness predicted by Teilhard de Chardin, McLuhan, and Thompson. This could certainly be compatible with, and supportive of, enhanced transportation on and off our planet. But a significant aspect of this new communication was oddly turned away from the real, outside world – offering a world purely of images, sounds, information – the world of cyberspace.

The appeal of interaction

Motion pictures in movie theaters actually made important contributions to the communication/transportation symbiosis. True, they were shown on screens, and were of course representations of reality – communication – not the real world itself. But movies showed people in motion on the screens. And, more important, viewers had to physically travel to the theater. In the first part of the twentieth century the motion picture theater was a key attraction on every main street of every American town. Even in the Bronx, where I grew up in the 1950s, the Loew's Paradise motion picture theater on the Grand Concourse was aptly considered a cathedral of cinema. It was stunning in its sweep, its staircases, its candy-counter art-deco decadence. It was well worth traveling to, whatever its movies. And most people felt the same about their local movie theaters. A friend who grew up in Italy in the 1950s told me that people would stand, two to three rows deep, craning their necks at the back of movie theaters in his home town. Visiting them was a social event in itself – a good occasion to get out of the house (and for teenagers, away from their parents), regardless of what was on the screen.

Television, on this reading, was a step backward, because it kept its viewers at home. Still, it freed radio for automobiles. In any case, like radio and motion pictures, television was a one-way medium. It was no more a convincing substitute for real life than books, cave paintings, and all the one-way media of history. Despite the criticism heaped upon TV by academics for decades – Lewis Mumford, Marie Wynn, Neil Postman, Herb Schiller most prominently among them – it was ultimately no more mistakable for reality than a cartoon. All except perhaps the youngest children knew that unlike people in the real world, folks inside the screen were not amenable to conversation. Or, as Socrates had noted millennia earlier about the published word, if you did try to talk to people in TV-land – if you asked Walter Cronkite or any news-anchor a question – you would get but one unvarying answer, what Walter was already saying.

Immediately interactive media, through which people could and did have conversations, had already demonstrated their pervasive power in the telephone. From its invention by Alexander Graham Bell in 1876, through the cellphones of today, the ring of telephone has been irresistible. It can pull us away from just about any aspect of life – be it sexual intercourse or a funeral – because the unknown caller could be fulfilling the personal or professional dream of a lifetime. (Caller ID only mitigates this slightly – we still have to give the ringing phone attention sufficient to allow us to look at the ID.) Letter-writing, of course, also had its interactive appeal, but the time lapse or asynchronicity between writing and reading and replying made it much less disruptive than the telephone – not to mention that written words are silent. And the recipient of the letter, in any case, was in the same boat as the Socratic questioner of written text or of Walter Cronkite on television: the letter could give but one unvarying answer to any questions put to it, namely what was already written in the letter.

The instant telephone thus exceeded by a universe even the most informal epistle. And yet for all its intrusive power, for all that the telephone did and still does conduct so many aspects of life, no one ever thought that the telephone was a *form* of life. No one talked about living in phone space. No one ever thought to call phone conversation a virtual or any other alternative form of reality. Everyone understood and understands that the telephone is a wonderful, maddening, gladdening, trivial, profound, annoying, life-saving lifeline in the world in which we already live. A pipeline through the world, not an another world in itself.

Not quite so the world of cyberspace.

The medium of media and the real world

John Perry Barlow famously defined cyberspace as the "place where people are when they are on the telephone." One reason that the telephone, prior to cyberspace, was not implicated in being a world is that, for all of the phone's astonishing interactivity, it provides information in just one form: sound. The Internet, in contrast, conveys information in every form known to every human medium in history – speech, pictures, writing, motion pictures, and so on. The Internet offers books, magazines, newspapers, movies, videos, radio, television. Indeed, it may be used as a telephone as well. This is why, in my previous book, *Digital McLuhan*, I termed the Internet "the medium of media." Other than the lack of physical presence – which lack is the subject of this book – the stuff of the Internet can comprise a world indeed.

The seeming totality of this service is what gives the Internet cachet as a "place" to be, a realm to inhabit, an alternate, "virtual" world. Looked at individually, the threads of the web are easy to connect to long-standing off-line human activities, and seem more like extensions of our world than elements of another. Buying books and all kinds of other items on *Amazon.com* is today's rendition of the Sears Roebuck and other mail order catalogs that were the talk of America the last time the century turned. E-mail is but the latest form of Victorian letter-writing. Chatrooms and instant, live messaging can be among the most radical departures from the past – and among the most conducive for giving the impression of entrance into another world – but even these are reminiscent of masquerade balls. And in the case of chatting online with someone you know, the departure is no more extreme than a walk in the park.

Of course, this park is worldwide – because the older modes are transformed online, and allow us to communicate in new ways. Radio online, via real audio and similar techniques, can be heard anywhere in the world, any time, unlike traditional broadcast radio, which requires listeners to be in the right place at the right time. And no ball ever brought together people from all over the Earth, instantly, and in a way that allowed children to masquerade as adults. But the roots of the real world nonetheless show through each of these transformations, and remind us that they are outgrowths rather than replacements of flesh-and-blood life.

Taken together, however, operating in concert, the components of the Internet point in a different direction. We can read our newspapers online; flirt online; buy graduation presents, trade stocks, bid in auctions, go to the movies online; even make love online after a fashion. It is easy to imagine, in that kind of environment with that kind of access to so many different kinds of activity, that we are someplace else – in an alternate world where we can move from one activity to the next in nearly no time. In this virtual world, whose only relevant laws of physics pertain to the speeds of data through wires and space, and the electricity that makes it all run, the physical location of the activity – where our online correspondent may live, where the website that makes our online movie available may be situated (in what physical computer) – is of little import. Once in cyberspace, all activities are equidistant, barring broken web links or other communication impediments or slow-downs.

The natural state of affairs in cyberspace – of pursuing any personal or business activity online – is thus equal, nearly instantaneous access to everything, which means that rapid switching from one activity to another, numerous times, is usually no problem (or, if difficulties do arise – say, in switching from research to flirting online – these stem from our attitudes about the activities, what we bring to them from our offline lives, and not from their locations). This is in marked, unprecedented contrast to our traditional realspace interactions, where the mere distance one must travel to keep an appointment decides whether the appointment should be made. Here, then, in the initiation and receipt of wide varieties of experience mutually reinforcing in their ease, the Web makes its most persuasive offer of another world.

Nonetheless, this offer is tested, if not shattered, not only when the real world rears its head and cuts off power, but any time a transaction in cyberspace requires realspace ratification or completion. A book purchased on *Amazon.com* still takes at least a day to arrive by overnight mail. True, it is likely that one day we will see more books available in e-text, and these can be conveyed instantly, in accordance with cyberspatial expectations, but that still does nothing for the online purchaser of a lawnmower.

Indeed, realspace speaks up, whatever we may be doing online, whenever we get hungry.

Two kinds of Java in the world

In a very real sense, information is not real: it is a representation of something else that is real. It has profound effects on real things. It is perceived and/or thought by real things – living things, human beings – either about themselves, or about other real things, or about other information which is a representation of something else that is real. Information certainly exists, and in that sense is part of the universe, and so in that sense is real. But the reality of information is very different from the reality of what it describes.

One proof that information is not fully real is the lie. Deceptive information describes things in ways that vary from how they really are; or perhaps a deception describes something as existing when it does not exist at all.

Another way we know that information is not completely real is how it fails, utterly, to have any effect on certain aspects of reality. Information about water may lead us to something that quenches our thirst. But if all that we had was that information and no water, we would die of thirst as surely as if we lacked that information in the first place.

Thus "Java" script – the programming language introduced in

the late 1990s to make images, words, and the content of the Web jump and otherwise move in interesting ways – could not move a real cup of Java one inch. Nor will any of Java's successors, as long as they remain descriptions of reality. As forms of information, they might well instruct machines to deliver coffee – just as spoken words might entice human beings to do the same. But the information on its own, in itself, can convey nothing except information – nothing real, nothing tangible, nothing literally capable of consumption. It would take a Java teleportation device – which disassembled the atoms of brewed coffee, shot them through a suitable conduit, and reassembled them into brewed coffee on the other end – to actually move a real cup of coffee anywhere. And if such a device is ever invented, it would be a mode of transport, not communication, for it would be trafficking in real things, not their representations.

Which is not to say that, because it falls short of reality, communication is always second best. It cannot satisfy physical hunger or thirst. It cannot, in the case of sex online, procreate our species. But virtual sex therefore contains an advantage for those who wish to have sex without babies. And virtual sex is also much better than a condom for protection from venereal disease. In that sense, sex online is not only virtual but virtuous.

Indeed, many species of communication about reality are better than the real thing – more uplifting, inspiring, more instructive about reality than the thing itself. Monet's painted water lilies teach us more about light and color than most of us would get from looking at the pond he saw, because Monet was able to subtly abstract and deftly highlight certain components or aspects of the pond and their interplay. Further, none of us can see anything of Monet's water lilies outside of their painting, because they decomposed long ago into their chemical constituents – here the painter, like Bazin's photographer, indeed rescued a delicate, fleeting scene, whatever different things Monet and we might have seen in it. But however more satisfying or instructive Monet's images of lilies may be, however better for the soul and the intellect, however much their blurry beauty may have defied death, they cannot be picked and placed in a dish of water, transplanted to another pond, cultivated to produce more lilies. On those scores, the virtual world is not virtuous but virginal.

But in the larger scheme of things this is not necessarily a problem either. The intrinsic, ineluctable inadequacies of communication do not matter as long as they do not preclude or impair or replace our access to the real world. There is time enough to enjoy Monet's watercolors as well as the colors of real water and lilies, as long as they are both available.

This is one of the key points of my argument. Cyberspace itself is not the problem, even though in many ways it is more alluring than books and television combined, because it is in effect a combination of those two, and many other media. But that miracle is not troublesome – unless it pre-empts or makes more difficult not just reading books and watching television, but walking to the bookstore, driving to the movie theater, strolling down the street, taking a plane or a train so we can stroll by the shore. Watching a movie about ships in space, moving at near speed of light on the Internet, is similarly no problem – in fact, both activities are great – unless they get in the way of our actually traveling in space at much slower speeds in the foreseeable future.

Ecologies of transport and communication

McLuhan talked about "sensory ratios," or the balance of seeing, hearing, touching, tasting, and (to a lesser degree) smelling, in our lives. McLuhan thought a preponderance of one or two modes at the expense of others provoked a powerful unconscious pursuit of the others, as if our psyches and bodies were striving to rectify the imbalance.

This way of understanding media can explain many developments. In the first part of the nineteenth century, the telegraph extended sight (reading words) across distance, while the photograph extended sight (literal images) across time. So in just these two inventions, speed (telegraph) and preservation (photography) were balanced, as were abstract (alphabet) and iconic (images) modes of seeing. But these media turned a deaf ear to sound – which cried out for its own, equivalent extension. It was less than 50 years in coming – not long by leisurely Victorian standards. In 1876, the telephone extended the human voice across distance, and a year later the phonograph did the same across time. Quite a balancing act.

McLuhan's favorite example of the provocation of unaddressed senses by a medium was television – the Internet of his generation (at least in the same sense that my baby-boomer generation sees the Internet: a medium that came to us after we were already well established as adults). The world of one-way seeing and hearing of 1960s television, McLuhan often noted, stoked a need to get in touch, to reach out and touch somebody. Among the early results were the "Summer of Love" in 1967 and the sexual revolution. A result that was later in coming (and which McLuhan did not live to see) was the Internet itself, which satisfied in other ways our need for two-way, interactive communication that unresponsive television had inflamed.

We could apply this notion of sensory ratios to communication and transportation in general, and expect that the growth of one at the expense of the other would be rectified sooner or later. Certainly we have the means at hand, here on planet Earth, to increase opportunities for physical transport. We can build better planes, cars, trains, scooters, highways, avenues – and in the case of trains (Amtrak's high-speed Acela) and perhaps scooters (the Segway), we already have. Indeed, the parts of our lives that entail physical presence seem too fundamental, in any case, to ever be entirely erased or eclipsed by communication, or even suppressed in favor of information transfer for any great length of time. In the 1970s, McLuhan argued that big cities were becoming obsolete, since the information they provided was available everywhere via television. This is even truer today in the age of the Internet. But the response I gave to McLuhan in the 1970s still holds: media may give us food for thought, but big cities will survive and thrive because they give us unique, real foods to actually eat. The city will endure, if only because of its restaurants. (I believe I told this to McLuhan as we dined in an excellent Greek restaurant in Manhattan.)

Big cities and restaurants seem to be doing fine these days – witness how well New York City rebounded after an attack far more deadly and destructive than anything an information technology, even deliberately turned against us, could ever do.

But the Internet does offer unique competition to transportation, and stands as a ready alternative to the improvement of planes and cars and trains, because of the Internet's easy interactivity and delivery of multiple services. In an age of terrorism and the understandable concern about air travel it instigates, video and web conferencing have increased at the expense of long-distance transport in the U.S.A.. Growth of meetings by information proxies – by images rather than real people – is a good thing, as long as it does not detract from the crucial development of better safety technologies and procedures for air travel. Increase of all options, giving people a choice of meeting by video or in person, is what we should be aiming for.

Movement into outer space, because it has barely happened as yet, is especially vulnerable. There are no restaurants there, no millions of years of physical human interaction to serve as a backdrop and network for further development. Like the cosmonauts in the Soviet Mir space station, who more than a decade ago looked down at the Earth as their country crumbled, and who wondered what life would be like for them if and when they returned, so American astronaut Frank Culberston looked down with his Russian colleagues from the International Space Station Alpha on September 11, 2001 at the plumes of deadly smoke rising from Manhattan, and then Washington on the next orbit, and wondered what the future would hold. Human presence in outer space is still too fragile, too new, too tentative to long survive calamities on Earth – especially when they afflict the major space-faring nation, the U.S.A. In subsequent chapters, we will explore what can be done to strengthen and make permanent our existence beyond this planet.

But first, in Chapter 4, we will look at a technology which, while increasing the invasion of sometimes unwanted communication into every facet of our lives, may also be seen as an heroic attempt to reclaim physical mobility in an age of information transfer, and indeed enabled acts of heroism on September 11: the cellphone.

4

THE CELLPHONE AS ANTIDOTE TO THE INTERNET

All technologies are trade-offs. Whether transport or communication, they disable, set back, and eclipse at the same time that
they extend, enhance, and provide. They cause new problems as
they solve old ones. The automobile was not as restful to its
driver as the train was to its passengers, and the automobile
brought exhaust pollution to every road. On the other hand, the
automobile enabled its drivers to come and go as they pleased,
on schedules of their own devising, and the putting to pasture of
horses as agents of transport eliminated piles of horse manure
and the tetanus bacilli they bred on city streets. Similarly, the
ring of the phone disrupted many a dinner conversation, as at
other times (or even at the same time) it conveyed heart-
warming, wonderful tidings.

Can we rightly consider these technologies as progress?

Yes, if their benefits outweigh their drawbacks, and if we can claim a net gain in human happiness. How many personal and business relationships have been well served, even made possible, by car and phone? How many lives have been saved? How many dreams have been realized, noble aspirations that would have been left in the gutters of older media?

The automobile is tougher to declare a clear winner on these criteria than the phone, since the deaths and injuries of car accidents must be laid at its doorstep. But an overwhelming majority of Americans apparently find the benefits of driving to far outweigh its risks.

The traditional telephone – the "landline" phone – is of course less deadly, indeed not deadly at all, but can be more annoying than the automobile. The cellphone ratchets up all parts of the equation, by being more helpful and potentially life-saving than the landline phone (we can receive calls anywhere), more annoying (we can receive calls anywhere – and others can hear them), and, in the case of drivers using cellphones, more deadly. (Since no sustainable evidence has yet been put forth to show a connection between cellphone use and brain cancer, I'll leave this out of the equation.)

So the cellphone is a trade-off *par excellence* – like all technologies, and, for that matter, all evolutionary systems (humans are smarter than great apes but physically weaker). Its main daily, actual benefit is clear and profound: we can keep in touch with those whom we love, or are doing business with, any time we desire. Its drawbacks are almost as clear and no less profound. Where the old-fashioned telephone shattered dinner conversations in the home, the cellphone intrudes upon conversations in restaurants, and thus disturbs total strangers.

Further, at its potential worst, the cellphone can literally shatter lives when the user is driving a car – which balances the cellphone's best possible benefit of allowing calls for assistance in life-and-death situations, in cars or otherwise.

But laws regulating such cellphone use show that we can rationally act to minimize the deficits of new technology. Contrary to critics of our technological age (such as Jacques Ellul in his classic *La Technique*, or Langdon Winner in his *Autonomous Technology*), who view us as hapless, helpless victims of our own inventions, we can take an active hand in refining their use and increasing their net benefit.

Sometimes the remedy can be legal. Other times the adjustment can be technological. The window shade allows people to control when the insides of their homes are viewable by the public, thus removing a disadvantage of the window (lack of privacy: the Peeping Tom) while maintaining its benefit (easy to look out at the world when desired). The VCR recovers television programs that would have been lost – after either a first viewing, or having not been viewed at all – just as the telephone answering machine can capture missed calls. These and other examples of what I call "remedial media" are, like laws, in effect medications for what ail us in our media, deliberately developed and applied to reduce, excise, or otherwise control a media problem.

But aspects of technological evolution are also frequently unintended. Certainly this is usually true of disadvantages created by technology. No one who helped market the first automobiles did so with the expectation that accidents involving these vehicles would kill and injure so many people. Nor did

American oil moguls such as Rockefeller, who encouraged the production of cars with combustible engines so that he could make millions of dollars, have any inkling that in the second half of the twentieth century this reliance on gasoline to move cars would empower nations with vast oil reserves in the Middle East.

In many cases, the main use of a new technology was unforeseen even by its very inventor. In the realm of media, Bell's initial impetus for the telephone was an attempt to improve communication for the deaf. Edison at first thought his phonograph would be a telephone answering machine (he was not ultimately wrong, just a hundred years premature). And he would make a similar kind of mistake in his initial vision of the motion picture process which he helped invent a decade later – he saw it as a natural accompaniment for musical recordings, which he had belatedly come to realize would be the primary use of his phonograph. As with the phonograph, Edison's expectations for motion pictures would come to be realized a hundred years later, in the music video.

If so many facets of media and their effects are unintended, we should not be surprised that a medium will sometimes have an unexpected, very specific remedial impact – that is, its amelioration of a media problem, its benefit, will be the result not of rational analysis, but of accident. Its curative effect will be an unintended consequence, just as highway deaths were of the automobile, except the consequence will help improve human life by reducing the negative impact of another medium.

That is just what the cellphone – in addition to all of its planned benefits and unintended drawbacks – does for the spatial segregation of the Internet.

Dissolving what glues us to the screen

For all that the Internet opened up new vistas in cyberspace, showing us things we could not possibly otherwise see from our homes or our places of business, it also insisted that we stay stuck in our seats and glued to our screens in realspace. That was the trade-off. If you wanted your mind to reach everywhere, your bodies were required to be in a very limited number of places, usually seated, indoors, in front of a screen. The brain soared and the backside suffered.

The Internet as a desktop medium is in fact more demanding than other media in requiring its users to stay in their place. Unlike television, which has the courtesy to continue when its viewers absent themselves to raid the refrigerator, visit the bathroom, or even doze – usually without much loss to their comprehension of the program, as long as the absence is not too long – the computer as an Internet browser usually does nothing without some continuing input from the user, and indeed can log off altogether when deprived of reassuring stimuli from the keyboard or mouse. Even when in front of the television and awake, the viewer can lean back, legs up on an ottoman or other convenient furniture, and relax. The Internet, in contrast, is a much more exacting provider – and recipient – of information. It insists on our almost complete attention – certainly if we are to read and write upon it – and that our faces be no more than a foot or two away. Ironically, faster browsers aggravate this problem; the less time it takes for web pages to load, the more frequent the ministrations needed from the user.

Books, of course, cannot go very far without hands to turn

their pages and eyes awake enough to read them. And we need to keep its pages close to our face. But books have the advantage of being readable anywhere there is light. In contrast, even radio, which since the invention of the transistor in the 1950s has been the most portable of electronic media, requires batteries. Books, certainly during daylight, come with batteries included and charged. And even in the evenings, electric lighting is practically ubiquitous, and in any case far more immediately available than batteries.

No current electronic communication device has the book's – especially the paperback's – almost effortless mobility, requiring no preparation (other than being literate and having the book at hand) to receive its information. Even the cellphone needs batteries, as all of us who have tried to use our cellphone with a "low-battery" indicator well know.

But the cellphone and its growing Internet capabilities is the biggest wireless leap thus far, utterly beyond both broadcast radio and the book in the interactivity of communication and the sheer diversity of information conveyed.

The integration of the Internet into cellphone technology, just beginning now with "Blackberries" and other wireless e-mail methods, but expanding to all areas of the web, may be the cellphone's greatest calling. The need to move personal computers from desk to desk, or off the desk entirely, was apparent from the very beginnings of the personal computer revolution. Kaypro's "transportable" CP/M computer in the early 1980s weighed 25 pounds, and packed neatly if inelegantly into a turtle-like case that was part of its console and screen. Radio Shack's M100 weighed only four pounds, ran on batteries (the

Kaypro did not), and became the world's first widespread laptop. But its internal modem (300 baud) required connection to a land-based phone system to access what existed of the Internet in those days, or any online system. Since text was almost all that awaited the online traveler back then, the M100 and other earlier laptops made ideal machines for writing and its transmission. By the mid-1990s, the profusion of images, sounds, and non-text media on the Web put much greater demands on the laptop. Palm-pilots would soon begin to fill that niche, and with connections to the Web that were wireless.

But by then the cellphone was already in far greater use as a simple conversation piece.

Talking again leads the way to walking, on this Earth

Wireless car phones were employed conspicuously by the very rich as early as the 1950s – see the 1954 movie *Sabrina*, in which the character played by Humphrey Bogart talks on his car phone. A half-century earlier, Marconi had telephonic aspirations for his wireless (which developed into radio instead). So why was the cellphone so long in coming as an everyday medium?

The road from idea to mainstream technology is no easy course. First, in order for the idea to become a technology at all, there must be means at hand to build it. Leonardo's schemes for helicopters were workable in principle, but the materials and methods needed to fashion his flying machines had not yet become available in the sixteenth century. Charles Babbage's plans for an "analytical engine" encountered similar problems some 300 years later; in a world without vacuum tubes, indeed with just the very first application of electricity in the telegraph, a serviceable prototype of the modern computer was incapable of construction. Historian of technology Roger Burlingame refers to these missing ingredients as essential "collateral technologies."

But absent hardware is not the only pitfall. For a technology to flourish it must not only be constructible, but acceptable to the world at large. People must be ready for it. Slow progress in outer space since the 1960s has certainly been due more to lack of public (or at least political) enthusiasm than hardware. We did not forget how to launch Saturn rockets to the Moon in the 1970s and after; we elected not to launch them.

Even the traditional telephone took seventy-five years to find its way into the homes of most Americans. Its slow start was due not only to the expense of creating a system of wires that reached everyone's home – a technological problem – but the conviction of many people in important places, including the President of Western Union in the 1880s, that the telephone would never be more than a "scientific toy." By the turn of that century, ten calls were being made for every telegram that was sent, but public perception of the telephone as an essential medium still lagged.

The story of videophones is even more revealing of the non-technological prerequisites of media evolution. A telephone with pictures was envisioned, and described to Americans, as early as *Tom Swift and his Photo Telephone* in 1914. Early experiments in television in the late 1920s were actually video telephones, which communicated pictures and sounds to two parties in a long-distance conversation. (Then Secretary of Commerce Herbert Hoover and President of AT&T Walter Gifford talked by videophone between Washington and New York in 1927 – the same year that Transitone introduced the first radios in automobiles. The Hoover–Gifford connection was referred to as "wire television.") Videophones – under various names including "picturephone" – were touted at just about every World's Fair in ensuing years. And yet even now, at the beginning of the twenty-first century, the videophone is in use mainly by overseas reporters and the occasional business executive, not by everyday people.

Why not?

The high expense of video cameras, admittedly a technological factor, was certainly an impediment. Transmissions only of voices required less complex equipment than transmission of voices and images. Even today, on the Web, full streaming video requires more computing power than equivalent audio. Whether analog or digital, communication of pictures and voices takes more doing than communication only of voices, which in turn takes more than just written words (analog media for communication of written words would be print, typewriters, handwriting).

But a powerful psychological concern no doubt has also obstructed the videophone. People like their privacy. For all that the telephone breached the castle walls of the home, and put us on call 24 hours a day, at least we could talk on the phone dripping wet, dressed slovenly or not at all, hair askew, unshaven, no make-up. In contrast, the videophone puts into public play not just our voice but our face and body, in whatever condition they may be. I suppose a "modesty frame" could help address this problem – a smiling photo of your face that your caller would see when you did not want to be seen live – but this would also call attention to why you did not want to be on the videophone in living color.

Despite its own incursions into privacy, the cellphone is not visually intrusive, and this is probably one of the reasons that it caught on more quickly (though by no means overnight). Mobile phones made their first significant commercial appearance in the 1950s, just as the fixed, land-based phone was becoming universal in the U.S.A. Mass adoption of cellphones awaited technological innovations ranging from telecommunication satellites to digital encoding. Not only were these in place by the mid-1990s, but so was a public chomping at the bit for a phone that could live in a pocket or a pocketbook.

Why was the world suddenly so keen for the cellphone? The overt reason was that it gave easier access to information. In Third World countries, with no infrastructure of telephone cables in the countryside, the cellphone enabled leap-frogging from the nineteenth straight to the twenty-first century in communication. In nations and societies that were already big players in the digital age, the cellphone provided another convenient avenue into that all-important, glittering domain which was first and foremost in everyone's minds: information.

But there was a covert reason as well – or at very least, a highly significant consequence of the cellphone that was the other side of the increased-access coin. For the cellphone brings people down to Earth, back to Earth, from virtual realms of digitally brokered reality. Jacques Ellul wrote in the 1950s how the technological age pulled humans out of their natural world and deposited them in cubicles in offices, like flies "stuck to fly-paper." However much the cellphone may intrude upon, dilute, impair our enjoyment of the natural world – or even the real world of sidewalks – in so doing it also puts us back into that world.

Looked at as this sly champion of transportation – one that comes heralded to us, and not incorrectly, as a champion of communication – the cellphone is surely a remedial medium for the problem of too much movement of information and not enough movement of people. In its sometimes irritating, nattering way, the cellphone nobly attempts to reassert the ancient balance of talking and walking.

If the cellphone heals part of the imbalance of communication and transportation in our lives, by reinserting the human into the world outside offices and homes, then the annoyance that the cellphone also engenders – by opening the gates to communication in the wrong places – can be seen as but the inevitable itch of new, healing tissue. Emerging customs and social consensus that distinguish appropriate from inappropriate use of cellphones are salve for that itch; laws that rightly prohibit drivers from distraction by cellphones are stronger medication. Eventually cellphones will be hardwired into just about every automobile, the same as radios, and voice-activated, so making calls won't take the driver's hands off the wheel or eyes off the road. Earpieces and personal microphones will be included for more private conversation. (These are increasingly in use already.) People will intuitively know when not to use cellphones in public, just as we know not to start reading a book or newspaper when we're in the middle of an in-person conversation – unless we want to insult our partner – and myriad other do's and don't's of interpersonal communication. The new tissue of the cellphone will become a fully functioning part of our system. It will do its job without itching.

But it won't heal everything. The cellphone gets us out of the office, out of the home, out into realspace on Earth but not beyond. This is because the cellphone is not a vehicle of transportation, but a vehicle of communication that facilitates transportation. Here on Earth, the cellphone gets us out into the world by allowing us to walk and ride without losing touch – without risking loss of communication. It hitches a ride on our feet, wheels, rails, roads, and wings. It adds value to means of transportation already available. It makes them safer, more useful, more connected.

But the means of getting out into space are barely available, if at all. And millions of people who would go out into space, if only they could keep in touch with folks back home via cellphone, are not among the factors that have hindered the space program.

In Chapter 5, we will look further into what has stalled our movement off of our planet, and why this new mode of transport is so crucial to our health as a thinking species.

5

THE ONLY WAY FORWARD FROM CALIFORNIA IS UP

What's keeping us down?

Why California?

It is the home of the Jet Propulsion Lab in Pasadena, true. But surely Cape Canaveral and Houston have better claims to the cutting edge of departures off of this planet, as did various places in the former Soviet Union in the 1950s and 1960s.

California's role in the quest for outerspace – for realspace beyond this planet – is as much metaphysical as physical. It is a state that also epitomizes a state of mind, and provides at least three crucial points of reference for our cosmic predicament of needing to leave this planet, and thus far succeeding only minimally.

To begin with, California is the end of the world. Not, one hopes, the end of its existence, but the conclusion of its physical extent, at least in the historical progression of the civilization currently in stewardship of this planet and responsible for the space age that has been dawning these past fifty years.

As far as we know, no mass migrations have gone from west to east, across the Pacific from Asia to North America, since the Bering Bridge was washed out some tens of millennia ago.

The Age of Discovery went the opposite way, first across the Atlantic Ocean, and then across the New World land mass. This expansion was a triumph of transport and communication – including the compass and the printing press in the first waves, the railroad and the telegraph (as well as the printing press, and its popularization of the West) in the last. Its terminus was the West Coast of the Western Hemisphere.

Call it California, for short.

Once our civilization arrived there, the only place genuinely new on this planet was up – or off of this planet entirely.

California's more thoughtful citizens in the second half of the twentieth century understood this – California is the only state in the U.S.A. ever to have had its own space program. Its progenitor and governor at the time was Jerry Brown. For this and other high offenses he was dubbed "Governor Moonbeam." Such is often the price of vision. (Marshall McLuhan – whose work Brown studied and took quite seriously, and whose daughter Mary served in the Brown administration – was also derided for his focus on myth and metaphor and symbolism, rather than down-to-Earth reality, in his explorations of media.)

But California was high in another way – one which ironically undercut its status as the logical launchpad to space. For this other way was a path not to outer but to inner space. And this other high – this psychedelic, LSD, Haight-Ashbury high – grabbed all the attention.

Thus, as Apollo voyaged to the Moon with people at the end of the 1960s, California celebrated voyages of the mind. Instead of improving the human position in the cosmos via technology, the drug culture sought to improve our perception. Unsurprisingly, the preoccupation with inner space in the psychedelic sixties had little lasting impact – other than distracting us from interactions with real cosmic reality, that is, from realspace. Perspectives and perceptions, however enlightened, go nowhere unless embodied in tangible technologies.

This applies not only to drug-induced excursions of the senses, but to a related exploration of inner space, one which was even more intrinsically Californian. Given its location on the edge of the Pacific, California's adoption of Asian paths to inner space – improvement and enlightenment via yoga and acupuncture – may have been only natural. Like psychedelics, it certainly caught on, and well beyond California. But also like mind-expanding drugs, and despite the somatic emphasis of tai-chi and its cousins – whose goals are to get practitioners in better touch with their bodies – the Pacific paths all led inside. Their pursuit attained the cosmos only in the microcosm of the intellect.

Cyberspace, the third Californian player in the struggle for space, bears significant resemblances to inner space. Both traffic in reflections, representations, and re-creations of reality. Timothy Leary gave each his endorsement, becoming an early apostle of cyberspace in the 1980s, just as he had been for

LSD in the 1960s. But Silicon Valley cyberspace also excels in social interactivity – economic, psychological, scientific – and this kind of reality testing sets cyberspace apart from the solipsism induced by drugs or meditation. Cyberspace thus sits half-way between realspace and psychedelic states of mind.

Mental states and underlying attitudes, whatever their sources, are crucial to the success of any human project. If an idea cannot get very far unaccompanied by technology, neither can a technology fly too long on wings unbuoyed by ideas. In this chapter, we consider some of the states of mind that have nearly dragged down our movement into space. If drugs and the quest for inner space were a distraction, these other attitudes have been an outright dead weight in our attempt to leave this planet.

Realpolitik versus realspace

Mesmerization by interior landscapes was not the only enemy of space exploration in the 1960s. The Cold War – the seeming antithesis of California dreaming – proved to be much more destructive to the long-term prospects for humans in space.

But was not the U.S. competition with the Soviet Union, played out on the precipice of all-out nuclear war in the aftermath of Hiroshima, the goading force for all successes in space in the 1950s and 1960s, from the Soviet Sputnik in 1957 through to the American walk on the Moon in 1969? Yes, it was.

And given the withering of human movement into space after 1969, it is hard to argue with such success – hard to see fault in anything that helped lift us off this planet, for the first time in our existence, as far as we know.

And yet a case could be made that the very Cold War which motivated the machinery of space exploration in the first place was also responsible for its subsequent self-destruction. To begin with, the psychology of science in the service of the Cold War was inimical to the usual standards and procedures of research. Science thrives on sharing of information, knowledge, and theories, so they can be tested by other scientists and refined. The Cold War mandated secrecy. Indeed, disinformation – the deliberate dissemination of false or misleading information – as well as sabotage were also tools of Cold Warriors on both sides. And although no outright evidence exists that sabotage played a role in the space race of the 1950s and 1960s, the goal of winning via better science would certainly have been served in the Cold War mindset by covert disfigurement of an opponent's work.

As the eventual loser of the space race, the Soviet Union was presumably the greater victim of any such disabling that did occur. Whatever the reasons for the Soviet failure to land people on the Moon (and return them) prior to the U.S.A., and the abandonment of any subsequent attempts, it remains one of the mysteries of our early Space Age. First to launch an artificial satellite, first to reach the Moon with any kind of space craft (Luna 2, 1959), first to put people in space and orbit, the Soviet Union fell behind to the extent that it reached the Moon with people not second, but never. The untimely death of Sergei Korolev in 1966 – the guiding genius of the early Soviet triumphs – is usually cited as a main cause of the Soviet inability to beat Apollo. The Soviet Cold War attitude that if it could not be first on the Moon then there was no point in going there at all – no propaganda payoff in being second – is usually cited as the main reason for the Soviet lack of interest after Apollo 11.

But the U.S.A. was a victim of this Cold War attitude as well, and in a way that fed off and reinforced the Soviet attitude – to the detriment and utter crash, in many respects, of the overall prospect of humans in space. If there was no advantage in being second to reach the Moon, neither was there much benefit – in terms of Cold War dividends – in going to the Moon again and again. So the U.S.A. also abandoned the Moon – for nearly three decades – after Apollo 17. A Soviet Union less in the grip of its own propaganda about being first or nothing at all – or perhaps less hindered by its opponent's covert activities – might have jumped on this opportunity and launched a massive second wave of people to the Moon. But Brezhnev and his successors were as numb to these possibilities as were Nixon and his. The glory days of the Cold War as a springboard to space would belong forever to Khrushchev and Kennedy. (Not Eisenhower – Kennedy's predecessor and U.S. President when Khrushchev and the Soviets launched Sputnik – who lambasted space travel as "pie in the sky.")

But the Cold War was by no means the only two-edged sword in the quest for space. Boosters that turned into burdens would continue in public attitudes about space, as NASA scrambled to sustain interest beyond the skies in the aftermath of Apollo.

The entropy of details and science fiction

One of the most depressing aspects of the space program in America after the 1960s was the attempt to justify it in terms of specific, practical benefits to folks down on Earth. Everything from Teflon to heat-resistant tiles were touted as unexpected bounty from new space technology. Although these claims were true, the game was bound to be a loser for space. Until we were far better situated in space, research and development and testing of new technologies would continue to be less expensive – and less dangerous – here on Earth.

That our movement into space needed to be justified by anything other than the cosmic wonder and profundity of getting out into the solar system, and then the stars, was the real problem. The Cold War had camouflaged the absence of this sense of wonder in the 1950s and 1960s. Or perhaps it existed in the 1950s, and withered in the 1960s – the ages of ecology (focus on Earth), Vietnam (an unpopular and unconstitutional war that led to dislike of anything military among many who opposed it), and getting high (inner rather than outer space).

The 1950s were also the Golden Age of science fiction, when Isaac Asimov, Robert Heinlein, and Arthur C. Clarke – the "big three" – hit their stride as novelists, along with several others. There was no shortage of wonder and excitement about space in the works of these authors, which made life beyond our planet seem as logical a next step in our development as a civilization as was movement from the Old World to the New. Indeed, space in these expectations and imaginations was the logical next evolutionary step for our species, as profound and portentous as our ascension from apes (shorthand for our immediate forebears in those less anthropologically sophisticated times).

Werner von Braun and other space pioneers readily admitted to being inspired by such science fiction and earlier works (in von Braun's case, going back to translations of other writers in 1930s Nazi Germany). But by the 1960s, a "new wave" of science fiction had emerged, one which sought to replace a sense of wonder with a combination of Joycean literary style and dark, cynical critique. This in turn begat "cyberpunk" in the 1980s – science fiction's handmaiden of the early age of cyberspace (a term which comes from William Gibson's cyberpunk classic *Neuromancer*, 1984). By the early 1990s, an editor of one of the three major remaining science fiction magazines remarked that no one was interested in space any more.

The success of *Star Trek* and *Star Wars*, of course, easily refuted such a claim, and science fiction on the screen certainly compensated for some of the slack of wonder in books and magazines. Indeed, science fiction in the movies and on television has no doubt been responsible for much of the continuing interest in space in the past four decades. But whether because of the greater intensity of intellectual involvement in reading, or the superior story-telling ability of the Golden Age authors, the cinematic science fiction of the past few decades has failed to inspire the kind of love and vision that drove space from Sputnik to Apollo.

In defense of science fiction, its dilution as an inspiration for space exploration was as much a symptom as a cause of the problem – a problem of inspiration for great scientific endeavors which had been fizzling out, like bubbles from an open bottle of champagne, for at least a century when it began to erode our move into space.

Philosophy versus science

There was a time – at least until about 150 years ago, to be more exact – when philosophy was king and science one of its vassals. There were actually three of these monarchs of the mind: natural philosophy (which would soon become "science"), moral philosophy (ethics, morality), and intellectual philosophy. The latter is what we today call, simply, philosophy. In a branch known as "ontology," philosophy inquires into the ultimate nature of existence – how things came to be. There is no shortage in sense of wonder here.

But practical science and invention are what captured the public's attention in the second half of the nineteenth century, and have stayed there, center stage – for better or worse – ever since. There was at first plenty of optimistic devotion to science and its applications, and with good cause. Diseases were being cured, people were traveling across and between continents in days or less, and communicating across these same distances instantly. But the World Wars gave body blows to the Victorian ideal of progress, and by the 1960s – the sunrise of the Space Age – a century's sparkling fascination with science had darkened. The victory of science over its parent philosophy was now a brooding thing in many thoughtful opinions, an inspiration streaked with dread.

Interestingly, science fiction, despite its name, was at its best a tableau for intellectual not natural philosophy – for ontology not science. Isaac Asimov's *Foundation* trilogy (1940s to 1950), for example, has next to no explicit science; travel between worlds occurs via unexplicated faster-than-light "hyperdrive." The real story Asimov tells here is about the effect that a hypothetical "science" which enabled a group to statistically chart the past and predict the future ("psychohistory") could have upon human existence. In other words, the *Foundation* stories are not about physics, astronomy, chemistry, or biology, but a process of organizing knowledge and information – they are more properly philosophy of science fiction or information science fiction than science fiction. They play more on Laplace's "Demon" – Marquis Pierre Simon de Laplace's early nineteenth-century notion that knowledge of all initial conditions permits perfect predictions of all future states – than any actual scientific discovery by Laplace or anyone else. Similarly, Asimov's robot stories (1940s to 1950s) tell us nothing about the specific construction of the "positronic" brains that he conjures up to guide these artificial entities. Instead, the stories probe to what degree the robots are human – and to what extent they can be controlled by their human creators. Much of the drama in these stories arises from the robotic capacity to do humans harm, even though robots are explicitly and convincingly programmed not to harm humans either by action or inaction. The drama thus arises in ethical, not scientific, quandaries.

Clarke's and Heinlein's works are also animated by themes that transcend science. Clarke's *Childhood's End* (1953) featured aliens from space coaxing humans towards our next evolutionary, cosmic leap – a theme Clarke would continue to explore in different venues, most famously in 2001, his 1969 movie collaboration with Stanley Kubrick. In all of these cases, Clarke conveys that the leap is at least as much metaphysical as physical. Heinlein's science fiction was the most explicitly military and gadget-oriented, but even his most memorable work explored humans struggling to make sense of laws of nature ripped open by new technology (for example, his *Time for the Stars* (1956), which tells the story of a twin who goes out into space and a twin who stays back on Earth, and how each age differently due to relativity effects).

This, then, was the "science" fiction that moved the first makers of space drives – a science fiction far closer to philosophy in its concerns, even religion, than to science in the laboratory. But by the 1960s, this Golden Age was replaced increasingly by a science fiction that owed more to Orwell and Huxley than to Asimov and Clarke.

Thus it was that as our ships left their moorings on Earth in the 1960s, they were also deprived of their deepest wellsprings of support – the quest to know who we are, what we are doing in this universe in which we find ourselves. Hyped up on the soap bubbles of the Cold War, justified in terms of scientific dividends which were either trivial or suspect, they drifted off-course, blown to ground by ill or irrelevant or distracting winds of drug use, ecology, and communication without transportation. Our first departures from the Earth were lost in space barely after they got there, because their guidance from Earth was occluded or gone.

There is a play – *Night of the Auk* written by Arch Oboler, debuting on Broadway in 1956, a year before Sputnik – which conveys the equivalent of the sad, true story of space. In the play, Earth's first ship lands on the Moon and the crew is jubilant; a national holiday has been declared back home. But before the crew can return in triumph for the celebration, their radio informs them that nuclear war has broken out on Earth. The artificial satellite that they were to dock in, around Earth's orbit, has been blown to smithereens. . . .

Atomic armageddon was a common theme in 1950s science fiction. It was all but retired in favor of ecological or epidemiological disasters by the 1990s (twenty-first-century terrorism has given it new relevance). But what made *Auk* (after the great auk, a bird driven to extinction by sailors in the 1840s) especially haunting was its vision of self-defeating humanity in the context of the cosmos – we cut our legs out from under us, knocked down the ladder, and the home below, just as we contrived to climb up to another celestial body for the very first time.

The play was not a success on Broadway – it closed after eight nights. I saw it on a television broadcast (*Play of the Week* – William Shatner portrayed one of the crew) in 1960. (A good rejoinder to critics of television, I should add, who say it is worthless or worse: How many people would have otherwise seen *Auk*, would even know about it? Not many, given its initial eight-day theatrical run.)

We did manage to reach the Moon in 1969, of course, and

avoid nuclear war on Earth. But the nightmare of *Auk* – of a few of our best selves stranded in space, watching, helpless, severed, as the rest of us on Earth destroy ourselves and our future – has never been very far from the surface. You could see its outlines shimmering in the Alpha Space Station crew looking down at Manhattan on September 11, as discussed in Chapter 3. We are vulnerable to this nightmare coming true as long as the roots of space travel are tangled and tenuous.

In ensuing chapters, we explore how our movement into space can be reconnected to its natural roots, and how that connection can be best nurtured.

6

FURTHER FROM HOME, CLOSER TO TRUTH

Marshall McLuhan thought the ecology movement – Earth Day, the *Whole Earth Catalog*, Gaia as an entity in need of explicit, full-time preservation – was ignited by Sputnik and its Soviet and American progeny, which for the first time gave us a view of our fragile, blue-green oasis of a planet. Certainly there were campaigns for conservation going back to Teddy Roosevelt. But the glimpse of Earth in its wispy, colorful entirety against the stark, black backdrop of space was something else. There we were, in all of our myriad life forms and civilization, trembling on the verge of existence, as John Dewey had said about us in a different context half a century earlier. Whether Rachel Carson had Sputnik and snowflake Earth in mind when she wrote *Silent Spring* (1962) – usually cited as the official first marker of the ecology age – is not known. But

Stewart Brand put a picture of the Whole Earth on the cover of his catalog. He said he had yearned for such a picture from the moment we first left this planet.

So we go into space to see not only what's out there, but back here. In the case of near-space, or space near Earth, "here" and "there" are almost in the same place. "There" is only a slight distance from "here," from the perspective of the universe, and so is an ideal spot to see us up close and in our entirety, in our majesty and insignificance, in our beauty and blemishes, in maximum clarity.

In the case of further outer space, or space further out, we see different things about us when we look back – we see more of the universe. Indeed, it is impossible to really see ourselves as part of the universe, or fully comprehend just where we are in the universe, from down here, or just out there, or even from several places further out there. We need multiple perspectives from all over the place, all over the universe, from myriad vantage points close at hand and way out there.

This was missed by most of the ecology movement, who looked only at Planet Earth, not at the space around it. This focus on the spotlight rather than the starlight, the figure rather than the ground, is a hazard every time we step out of ourselves and look back. It starts when we look in the mirror, and see our smile, our face, our clothing, and not the room in which we are standing. In the case of Planet Earth, it can disguise some of the most important insights which a view from afar can hold for ourselves and our future.

Mirrors: pitfalls and opportunities

McLuhan was also fond of talking about mirrors, in particular their anaesthetizing effect on eyes that look into them. He saw the Greek myth of Narcissus – the vain, handsome youth who fell in love with his own reflection in the water, thinking it was someone else, and died pining for it on the shore – as symptomatic of our age. It certainly describes aspects of the digital fascination with information and image over immersion in physical reality – a state of affairs that emerged after McLuhan's death in 1980 – and the shallow lessons we have so far gathered about Earth from our limited positions off-planet.

But used properly, a mirror can be an eye-opener. Everything else that we know about our appearance is second-hand, indirect, what we can glean from the expressions and behavior and words of others who interact with us. A mirror-image is first-hand – it shows us almost as we are, almost as we are seen by others. The reflection of reality is imperfect, not only because it is reversed, but because we are aware that we are looking into a mirror, and adjust our mien accordingly. But it is nonetheless an extraordinary improvement over no image at all.

The first person to look upon a pond and realize what was living in the light on the surface – what was thinking, and reflecting on the reflecting – changed the course of human history, perhaps even helped create it. This would have been an act of pure understanding, requiring no technology or invention. Presumably this act occurred – once or a million times – prior to cave paintings and writing, which as modes of reflection may be seen as the pond-reflection's progeny. The epiphany of the pool probably took place before the use of shiny, portable surfaces for mirrors, and certainly before photography (which produces literal, not mirror, images) and its motion, video, and digital offspring spun the image and the world around faster, slower, in all kinds of angles and directions in the past century and a half. Seen in this light, the reflection in the pond is one of the parents or grandparents of all subsequent media, second only to speech itself in importance to humanity. On that score, the human face that gazed into the pond did not mistake the face for someone else's and then die pining on the shore like Narcissus. Nor did the human gazer jump into the pond in an attempt to embrace the shimmering image, and drown.

Instead, the mirror and its children have been by and large revelatory. The whole enterprise of "objective" science stems from the mirror, and the recognition that we must have some distance from what we seek to understand. McLuhan, again, had some thoughts on the subject. He held the amphibian to be a much better analyst of water than the fish, since the amphibian could experience not only the water but where the water ended – could be in and out of the liquid box. The advances in psychology, once it moved beyond introspection, in all sciences once photography provided an enduring record of ephemeral events, once it stole time from the world, attest to the amphibian advantage. Of course, no scientist or science can ever be totally objective. Quantum mechanics suggests that the very observation of subatomic particles has impact on their properties. But these are, at most, places in which the mirror does not work – places where light itself, the very electricity of the mirror, does not behave as expected. Distortions in one part of the mirror, or when the mirror reflects an unusual subject, do not negate the clarity it can bring to other areas, or our ability to improve the distorted image by recourse to other kinds of mirror.

Among the important mirror-tools of science – literally and figuratively – are the telescope and the microscope. One looks out at the stars, the other looks back into us. The contribution of the telescope to our understanding of the cosmos and our movement into space is obvious and enormous. Here we consider some of its disadvantages, and the importance of keeping a special kind of microscope at hand.

Telescopes into microscopes

Opponents of the telescope are easy to ridicule in history. Most infamously, some Church fathers derided Galileo's telescopic observations – and the support he therein adduced for Copernican cosmology – on advice given by Aristotle, two millennia earlier, that the testimony of our naked senses should always be trusted over evidence provided by technological contrivances. This probably made sense in Aristotle's day, given the level of lenscraft then. It manifestly did not in Galileo's. Further, the proposition that the telescope was inferior to naked vision as witness of the Moon could be refuted any time on Earth, by training the telescope on some spot far away and then traveling to it, to corroborate or refute the observation.

But was Aristotle's advice wrong in every respect? One benefit of naked in contrast to telescopic vision is that things we observe with our unmediated eyes may be touched, sifted, sniffed – that is, examined by other, immediate modes of human perception. In the absence of such collaborative faculties, we are prone to make assumptions about what we see via telescope far away. For example, red shifts in observed objects in the cosmos were presumed to be indications that the object was rapidly moving away from us. Other explanations were then offered for the red shift. The point is that, until we can actually approach the questionable object directly, with multiple modes of perception, our best explanations and identifications are hypothetical conjectures. And, usually, the further away, the more hypothetical. (One could argue – as per the philosopher Karl Popper – that all observations and theories are conjectures, including those down here on Earth, since all of those are subject to error. But those amenable to less observation are more susceptible.)

Auguste Comte's statement more than 150 years ago that we will never know the chemical composition of stars, because how could we ever get there to perform the necessary tests, is cited frequently (including by me) as evidence of technological myopia: the science of spectroscopy, which gave us readings of starry composition by analysis of their light conveyed by telescopes, soon proved Comte wrong. But how wrong? Until we are able actually to travel to the stars, spectroscopy applied to their composition will lack corroborative support available for objects closer at hand, such as planets in our solar system. The enduring error of Comte may be not in his insistence on first-hand analysis, but his assumption that we will never be in a position to conduct it – not that he overlooked what telescopes could do with light, but what spaceships could do with tangible reality.

Recent suggestions that analysis of Martian soil scooped up by U.S. Viking craft in the 1970s may have overlooked evidence of bacteria brings home the importance of hands-on examination of distant objects. No amount of telescopic observations of Mars, no technologies that convey only representations, will accomplish what a few bits of soil can do in the possession of scientists with appropriate technology.

Further, we may conclude, after re-examination of the data obtained in the 1970s, that bacteria were indeed on Mars. But without continuing access to the actual soil, which would enable us to re-conduct and expand the testing, will we be sure? This, of course, is not an argument against the use of telescopes. Rather, it is an argument against the use of telescopes *only*. They are crucial. But they are not enough. They work best as scouts of objects and phenomena later susceptible to our direct, physical scrutiny.

There is obviously one planet where we already have such physical access – continuously, twenty-four hours a day. Planet Earth. Telescopes that look back at us on Earth are thus more like microscopes, since they examine in a different light, from a different level, what we already have all around us.

Indeed, inasmuch as a microscope is intrinsically a device which looks at something physically at hand – but not graspable as such in our hands (bacteria, viruses, molecules, and atoms cannot be meaningfully held in our hands) – we might say that what we need in our observation of far-off places in the solar system, galaxy, universe, are telescopes transformed into microscopes. We can no more hold a planet than a subatomic particle in our hand, but we can be physically involved in the planet's examination.

Apartment, city, universe

Our inability to comprehend much of our position in the universe, galaxy, solar system from a vantage point on or close to Earth should be obvious. Only an argument that such comprehension is unimportant – absurd (at least, to me) on its face – could justify, philosophically, a lack of interest in getting far away from Earth for the purpose of improving our perspective.

Literature abounds with depictions of the distortions and limitations of the cloistered viewpoint. Plato's allegory of the cave is probably the most famous. He held that the human capacity to understand only the superficial outlines of things, not their true, inner core, was akin to people sitting in a cave, backs to the daylight, able to see only the shadows cast by the reality outside on the cave walls within. Plato, pessimist that he was, saw this jail-cell of flickering distortions as utterly unbreachable (except perhaps, and then even only partially, by brilliant, right-thinking philosophers). The denizens of the cave were hopelessly its prisoners, much as citizens in more scientific millennia were (and are still) thought to be inescapably in the grip of entropy. I disagree on the inescapability, in all sectors. Regarding entropy, why accept the view that the universe, and everything in it including life and intelligence, is an infinite down escalator, rendering futile even the most heroic attempts to run up? If we agree that life and intelligence have anti-entropic qualities, why not hold out the possibility that human technological intelligence can eventually figure out a way of reversing the down escalator? (This is indeed the perspective suggested by a few optimistic philosophers and cosmologists, including Buckminster Fuller and Freeman Dyson; I also develop the position in my 1988 *Mind at Large: Knowing in the Technological Age.*)

Meanwhile, regarding knowledge of the universe beyond Planet Earth, the remedy to the Platonic problem is far less metaphysical: all we need to do is physically travel to the objects depicted in our telescopic shadows. The cave of Earth permits egress by spaceship, as has been demonstrated since Sputnik.

In any case, Plato's allegory could stand some space-age updating. Let us begin with someone standing near a window in an apartment in a big city.

How would someone confined to an apartment all of his or her life know about the city? What would the eternal apartment dweller know of street lights, traffic-lights, traffic patterns, parks, restaurants, and department stores? If we wanted to be strictly Platonic, only what shadows sliding through the window revealed on the walls of the apartment. But let us be a little less literal in our translation of Plato's allegory, and allow our apartment dweller to turn around and look out of the window – what would be seen then?

Only, at first, what was coincidentally visible from the window of the apartment – whatever traffic-lights, restaurants, department stores, and so forth just happened to be in window range. Given the teeming immensity of what we (with our God's-eye view in this allegory) know to actually be in the city, looking directly through the window is scarcely better than looking at shadows on the wall: both views are extraordinarily narrow, and guaranteed to be unrepresentative of the city as a whole.

But let us allow that, after x number of years, centuries, millennia – for in this allegory our apartment dweller stands for our entire species, throughout time until now – our denizen of the apartment develops the equivalent of powerful telescopes that could look out upon other parts of the city, places not visible from the naked-eye window view. No doubt our denizen's knowledge of the city would increase. But how far? With what depth?

Let us assume that there were no restaurants in the naked-eye window view. Would the telescope reveal them? Only if the eating establishment were in the telescope's line of sight. OK, let us assume that one was – a pizza place, visible via telescope pointed down a long, wide avenue, miles and miles of it stretching forth outside the window. Would the apartment dweller comprehend fully the pizzeria's function on the avenue? Well, people would be seen entering the place, and exiting with slices of pizza in various stages of consumption. So far, so good. But how much did they cost? Why did a few of the slices look white rather red-and-white? And what were those big, flat cardboard boxes that a few of the customers seemed to be carrying out? Unless the pizza place did at least some of its business on the sidewalk, with a menu visible on the street, most of these subtleties would be lost to the distant telescopic observer. The only chance our "scientist" in the apartment would have of fully understanding what was going on in and around the pizza place would be if he or she had already favored such places, in the past, with in-person patronage – had traveled to other planets and stars, in the reality underlying this allegory.

And the above is a best-case scenario. The windows of the apartment might well open up to a back alley, or only to a beautiful Hudson River view, affording no glimpse of restaurants, or traffic lights, at all.

But let us see if we can break out of the window-pane in a different way. One source of possible assistance to our inquisitive viewer might be dwellers in other apartments, either in different locations in the same building as our original observer, or, better, in buildings in different parts of the city. If we allow that our apartment dwellers invented not only telescopes but telephones, then their information pooled across various locations in the city could provide an accurate description of pizza places and their function.

But if our initial apartment stands for Planet Earth in this allegory, then our apartment dweller would be unavailed here too. For there are no humans, as far as we know, living on other planets or (with the notable exception of the International Space Station – certainly a step in the right direction) anywhere other than on Earth. We are all of us in the same boat, here – the same apartment – at this point.

Thus our predicament in the single apartment/planet seems intractable, as long we are obliged to stay in those confines. And it leaves us vulnerable, with no direct knowledge of food shortages and supply problems, water-main and power outages down the street or across town, that could greatly inconvenience or even kill us. We might even see reports of these in the mass media – let us allow television and the Internet in the apartment to stand for a different kind of far-gazing telescope for Planet Earth, say, the Hubble – but without the wherewithal to leave our apartment/planet, what good would these media reports do us? Even if we knew that something out there might interfere with our food/energy supply, where else would we go to get more?

In contrast, information about shortages and other dangers on Planet Earth is extremely valuable, because we are in a position to act on the basis of such information. (The equivalent for our apartment allegory would be information about something happening within the apartment.) This distinction between information that can be acted upon and information that cannot is crucial, and gets at the nub of the problem of cyberspace without realspace: information, of which cyberspace is a highly developed example, loses value to the extent that it is not actionable.

This does not mean that information which cannot be acted upon is useless. Sometime in the future it may find us in a situation where we can put it to practical use. Even if not, information may amuse us, and this is of value. (Information that inspires us is practical if it inspires us to do something.) But we do need to be concerned about developments which increase the ratio of non-implementable to implementable information. This is our state of affairs in outer space, every time we see something new that we cannot travel to.

Ultimately, lack of first-hand knowledge of where our world fits in the universe can become lack of life. This pertains not only to our sun going nova – which even I admit is so far in our future as to not warrant imminent agitation – or to our planet being taken out by an asteroid. The greater danger resides in what we now know nothing about – either because we see nothing, or have no or insufficient understanding of what we see – the myriad possible calamities that may escape our notice because we simply are not in a position to take such note.

And even where we do have an inkling, as with our possible collision with some asteroid, our lack of mobility makes us scarcely better than the eternal apartment dweller confronting a fire, or the amoeba and its ancient one-track way of dealing with dangers. We may see and hear far-off perils with our technological eyes and ears. But such advance warnings are of limited value if we cannot get out of harm's way. Just as the amoeba, blind and deaf, dies with the noxious objects that touch it, so will we, stuck on this world with all of our telescopes and televisions and Web connections, unless we are lucky and fast enough to devise a means of destroying the threat before it reaches us. (Our instruments of observation and communication would of course be crucial in devising and applying a successful response to the asteroid assassin. But even that, again, is predicated upon our recognition of the threat – we may not have that asteroid advantage in combatting other potential celestial calamities.)

In the end, then, the deficits of life confined to an apartment stem from a lack of transportation – or the imbalance of transport and communication, of information moved far further and easier than people, that in fact now characterizes our existence on Earth. Except that here on Earth, we can move, at least some of the time, in response to danger – we can take action if we see something bad on the horizon. But without far superior modes of space travel than we currently possess, we cannot do the same regarding threats and destinations off-planet.

Continued influx of information about events off-planet can certainly help – the thrust of this argument about too much communication and too little transportation is not that we should shut down our telescopes. Clearly a situation of some information and next-to-no transportation is preferable to next-to-no transportation *or* information. But in the long run, information alone won't be enough.

Robotic emissaries offer interesting opportunities on this analysis. They can certainly perform direct inspections and interact with alien landscapes and objects. They can send samples back to us on Earth, which offer a lot more than remote observation. Yet they cannot perceive the world – any world – with our human sensibilities (unless they were made to be so close to human that they were, by any reasonable standard, human). Robots and such issues deserve their own chapter.

But first, having considered some of the tools and perspectives that may clarify and assist our inherent relationship to space, we will look more carefully at what social systems are most likely to help us get there in Chapters 7 and 8.

7

IS DEMOCRACY THE BEST LAUNCHPAD TO SPACE?

Here, in political context, is a short history of our progress into space. The vehicles that would lift us off this planet and bring us into space were first launched by the most vicious, genocidal, totalitarian regime of the twentieth century, possibly in all of history: Nazi Germany. The first artificial satellites, the first animals, the first people, were launched into space by another totalitarian regime which, while not explicitly genocidal, was responsible in some accounts for more deaths in the 1930s and 1940s than Nazi Germany: this was the Soviet Union. That nation was also responsible for our first physical encounters with the Moon. (They were called "unmanned" at the time; let us refer to them as "unpersonned.") Finally, the actual landing of human beings on the Moon, and their safe return, was accomplished by the world's leading democracy, the U.S.A. In the years following, the U.S.A. reached Mars robotically and made advances in space-shuttle technology, while the Soviet Union, before it collapsed at the end of the 1980s, was a pioneer in space stations and human endurance in space.

What does this tell us about democracy as a political launch-pad to space?

Well, the signals are mixed.

On the one hand, we – the democracies – survived. The U.S.A. and its democratic allies (along with the totalitarian Soviet Union) beat Nazi Germany. And then the U.S.A. and its democratic allies triumphed over the Soviet Union in the Cold War. If we are talking about long-term support of humans in space, the perishability of totalitarian systems – at least in the twentieth century – cannot recommend them.

But there is the other end of the process, the crucial begin-nings of endeavors. And in the initiation of human movement into space, the democratic U.S.A. comes off poorly. Indeed, of the two horror-weapons unleashed at the end of the Second World War – the V-2 rocket by Nazi Germany, the atom bomb by the U.S.A. – it is the rocket that has been our conveyance to space thus far.

In this chapter, we sift through the historical evidence of var-ious political systems and their support of major technological endeavors such as space exploration. If we find that democracy is worse than the least worst system – to paraphrase Churchill – this will not be a clarion call to space enthusiasts to endorse their local totalitarian parties. Rather, it will be a stimulus to do the best we can with our democracy, where recognition of its draw-backs is a necessary step to using the system more efficiently as a ticket to space.

War and peace, application and invention

The totalitarian and democratic triumphs in space are united by an ominous thread: all were accomplished in wartime, or quasi-wartime. The V-2 was produced by the Second World War, and everything else from Sputnik to Apollo by the Cold War. Does this mean that our greatest leaps in technology are dependent upon wartime or near-wartime mentality?

Not quite.

Thomas Edison, a veritable one-man age of invention, created the phonograph, electric lighting, motion pictures, and dozens of lesser devices such as the pencil sharpener in peacetime conditions. So did Bell for his telephone, Marconi for his radio, the Wrights for their flying machine, and, three-quarters of a century later, Jobs and Wozniak for their personal computer. True, the airplane's subsequent development was helped enormously by twentieth-century wars. But the automobile did just fine under the peacetime ministrations of Henry Ford and others. The upshot seems to be that war is certainly not necessary for initial invention, and likely not as essential as we might think at first glance for the application and development of technology, either.

But is space travel just another technology? Or is the prospect of moving some of humanity off this planet in a reliable way so extraordinary that it defies useful comparison with all other technological episodes and sagas?

The key psychological edge provided by war is that everyone in society, including those who develop technology, correctly see their work as ultimately or even imminently a matter of life and death. This distinguishes war mentality from even the most cut-throat business state of mind: the entrepreneur, the corporate chief executive, the financier can be reasonably sure of living, perhaps even living well, if the business goes under. Failure, though unwanted, is always an option in business. In war, it cannot be – certainly not in the all-out, citizen-involving, ungentlemanly wars of the twentieth century, nor in the new kind of anti-terrorism war of the twenty-first century. And, since nuclear annihilation was an ever-present possibility had the Cold War heated up – indeed, mutually assured destruction (MAD) was held to be one of the main factors that prevented it from heating up – complete failure was not an option for most of the Cold War, either. In the end, the Soviet Union disintegrated from the inside, not from outside attack, even though the result could rightly be called a victory for the democratic U.S.A., at least by default.

This life-and-death mentality of war and near-wars, in contrast to business, in a sense injects totalitarian attitudes into democratic societies. In the interests of winning – or preventing defeat – young people can be drafted against their will (a major issue during the Vietnam War, since North Vietnam never attacked America), people of any age can be told where to work, food and energy sources are rationed, and so on. In our terrorist age, the U.S. government has begun to examine private e-mail. This makes sense when innocent lives are literally at stake. But if this is the price democracies would have to pay to provide sustained support for space exploration in peacetime, is it too much?

The challenge thus resides in making democracy a more

reliable foundation for space, without deforming it into a total-itarian-like system.

Microsoft space?

Karl Popper convincingly showed in *The Open Society and Its Enemies* (1945) how Plato's philosopher-king – the wisest in soci-ety having dictatorial power – leads to the Nazi party and the Soviet politburo. This is bad news for space exploration, which would be well served by a benevolent philosopher-king who committed resources sufficient to disseminate human civilization beyond our planet and eventually our solar system. But even in ancient Rome, the philosopher-emperor Marcus Aurelius was succeeded by the disastrous Commodus. Philosopher-kings are pipe-dreams that turn into nightmares. Their saving grace, cer-tainly in the modern age, is that, unlike democracies, they do not seem to last very long. This would make them unsuitable for support of long-range space efforts in any case.

But democracies, alas, are also unreliable – not only as escala-tors to space, but as supporters of any technological endeavor. The U.S. Congress not only slashed funding for space after Apollo, but failed to provide any funding for the Supersonic Transport Plane (the SST) in the late 1960s, and barely enough to keep the national rail system functional via Amtrak. (Again, in times of war, the U.S. Congress acts differently – it quickly pro-vided 17 billion dollars in funding to help airlines stay in business after September 11.)

Indeed, the U.S. government not only often fails to support

technological innovation, it often, in its anti-monopolistic zeal, takes technological innovators to court – as it did with Microsoft.

Monopolies arise from philosopher-king enclaves in democracies. Indeed, all businesses in capitalist societies have aspects of the philosopher-king, or the benevolent dictatorship, about them – certainly entrepreneurs and privately owned companies, but even publicly held corporations are inherently less democratic than a democratic society at large. This is not surprising. There are many non-democratic systems in a free, democratic society, some of which are even essential to the health of the democracy. These include families (where parents quite appropriately have dictatorial power – or at least try to – in some matters), classrooms (where the teacher not the student decides upon and grades the assignment), religious groups (where revelation is to be followed not contested), and even U.S. Presidential cabinets (where one vote, the President's, counts more than every other cabinet member's vote to the contrary, as per the anecdote of Abraham Lincoln and his cabinet – Lincoln polled his cabinet for a decision, everyone voted "No," Lincoln concluded by voting "Aye," and added, "the Aye has it"). And a business leader or board can usually allocate funds and resources far more efficiently than any democratic government.

But business success – survival – is another matter. Businesses (and their quasi-dictators) go under a lot more frequently than democracies, which is a good thing for freedom (that is, it's good that democracies have a better rate of survival). And yet, ironically, if a business succeeds spectacularly, this can put it on a collision course with the government. In its best light, the government's dismantling or even restriction of a highly successful business may be seen as a move to prevent the philosopher-king corporation from in effect enslaving the overall democratic society. In its worst light, the anti-monopoly campaign can be a myopic punishment of success – a success uniquely beneficial to society.

But to what degree has business spearheaded major, transformative technological growth in the last two centuries? In the U.S.A., the railroads in the nineteenth century, and telegraph and telephone in the nineteenth and twentieth centuries, would be prime examples. The laying of track and stringing of wires required decades upon decades of sustained work. The automobile may also be considered a triumph of industry rather than government, though the Interstate Highway System, initiated by the Eisenhower administration in the 1950s, has been a major Federal accomplishment. Radio, television, and air travel have been mixtures of business and careful government regulation from almost the beginning. And then there's the personal computer. Initiated by the entrepreneurial Apple, led briefly by the traditionally corporate IBM, since the mid-1980s the digital revolution has been spearheaded by Microsoft, and its wildly successful software ranging from MS-DOS to Windows.

The U.S. government's attempt to dismantle Microsoft, or at least slow it down, hit an impasse in the courts, which did not buy much of the government's claim that Microsoft was preventing rivals from marketing software that might compete with Windows and its allied programs. (I thought such a claim was riddled with logical inconsistencies on the face of it. For example, it was alleged, at one point, that Microsoft was packaging its programs so that users could not use Netscape Navigator, a competitor to the Microsoft Explorer web browser. But Netscape arose in the mid-1990s – precisely the time when Microsoft was supposed to be wielding all of its massive, monopolistic power. If it was so powerful, then how did a start-up like Netscape arise in the first place?) The out-of-court settlement between Microsoft and the government looks like it will leave Microsoft not only intact but retaining most of its power.

Let us assume, then, to make our inquiry into Microsoft as a suitable springboard for space as clear-cut as possible, that either the government never launched an anti-monopoly suit against Microsoft in the first place, or that Microsoft will indeed prevail in a way that siphons off little or none of its strength. What might we expect of an unmitigated Microsoft as a vehicle for our movement into space?

Well, the company certainly has the money – if not as much as the U.S. government, certainly enough to support a variety of endeavors into space. But it has a serious disadvantage, money-wise, in comparison to any government: it could run out of money. Governments (certainly local governments) occasionally run out of money too. But governments can go into debt a lot deeper and longer than any business. And governments have the option of raising money by increasing taxes. Businesses can make money only via successful investments, or selling things.

This lays bare the Achilles heel of business as a launchpad to space. Not only does business have to make money somewhere by selling something – the money needed to support space work – but what exactly can it sell in space? The railroads began by selling transport from one populated place – usually a city –

to another. Expansion into sparsely settled areas, which ignited population movement and growth there, came later. (And these areas were at least settled, albeit sparsely – unlike outer space, which, as far as we know, isn't settled at all. Indeed, no area on Earth is in principle unsettled or unpopulated in comparison to areas off of Earth, since any area on Earth is reachable by people sooner or later, via some means of transport. On Earth, the question is: How long does it take to visit your nearest neighbor? Off Earth, there are no human neighbors as yet.) The telephone began and grew the same way, selling conversation to people who were already able and willing and eager to talk. There was no point in having a telephone with no one to call or call you, so the telephone first caught on in densely populated urban areas, where short lines were easily strung and an individual was most likely to have someone to talk to on the other end, either for business or pleasure. In the long run, the telephone cashed in on the inherent inclination of people, everywhere, to talk. As a vehicle of conversation, the telephone sooner or later sold itself.

The personal computer was a little more difficult, since its use required some learning on the part of the population. But even so, its immediate benefits were obvious to the writer who put aside the typewriter for the word processor and its ease of correction, the work-at-home parent who could now connect to the office via computer, and a host of other professions. Further, the consumer niches for what the personal computer and then the Web could do were already at hand. Apple, IBM, Microsoft, and their competitors could see a way to make lots of money – selling devices and software that could be used to send mail instantly, anywhere else in the world with a connected computer, deliver news instantly and globally, buy books, auto-mobiles, clothing, just about anything, over the same distances. The subsequent failure of many "dot com" businesses at the turn of the millennium showed there was a limit as to how fast and far online business could grow. (And, as has been argued through-out this book, part of that limit stems from an irreducible need for realspace, face-to-face interaction, including in commerce.) But that there are real, permanent niches for computer com-merce – niches that build upon established needs that are hundreds, thousands of years or more in existence – has never been in doubt. Indeed, in the case of Web life, the very in-person patterns that serve to limit the growth of online business are the in-person patterns that inspire and, in a psychological sense, sustain the online world, by making its activities – such as purchase of clothing and books – familiar and obviously worth-while even though the online process is new.

Where, then, are the money-making niches in space?

Tourism – sightseeing the Moon, the planets, and the stars from a vantage point off of this world – is often cited as a logi-cally commercial venture for space. The unique cosmic beauty of the scenery is undeniable, unavailable anywhere on Earth. But that very absence from our planet might well discourage customers. The Everglades, the whaling waters off Cape Cod, even the Himalayas are all a relative stone's throw from civiliza-tion and its comforts – a small world-of-difference closer than anything out in space. Indeed, the immense distance from home is presumably part of what would make the trip beyond Earth appealing to the tourist in the first place. But how many people would really risk a journey measured in years, the time it would take for a round-trip excursion to any place in our solar system other than the Moon?

Outer space as a sightseeing industry thus would have to start very slowly and carefully. For a long time, it would no doubt be a plaything of the very rich. This could ultimately work, and support some movement into space. In decades and centuries ahead, space stations and settlements on planets, and faster travel speeds, could make the journey more appealing. But the trajectory of such growth would be utterly unlike the lightning of Web development that drew in tens of millions on a yearly basis in the late 1990s. It would likely be much slower than even the telephone, which took some seventy-five years to reach more than 50 percent of America's population. The upshot is that tourism in space will help over the long haul, but will likely be of little assistance in our current predicament – our need to jump-start a more regular progression of people off this planet.

Are there any other ventures in space likely to attract and support a sustained, massive influx of capital? Mining for resources on asteroids, construction of off-planet processing plants which could operate more efficiently than on Earth in low or no gravity, are popular and not unreasonable candidates. But the analogies to equivalent successes on Earth in the past are stretched here as well. The California and Yukon gold-rushes featured easy access to food and water, and free oxygen everywhere. A financially efficient processing or production facility in space presumes either robots that do not require food and water and air, or reliable supply of these to human workers, and in any case a dependable way of getting the results of such labor back to Earth. All of these are very likely in the future once we have already established ourselves in space. None of these are available at present, in our Earthbound culture and business.

On balance, then, business is not likely to exceed busted totalitarian governments and unreliable democracy as a spearhead into space. If Microsoft stays wealthy and powerful enough, or some other corporation achieves its success, it might be willing to risk the requisite capital for sustained space activities. Or perhaps a consortium of powerful companies could help. This would be in pursuit, at first, of public relations and possibly more profound cosmic benefits (if those in charge of the business were of such a mind), rather than direct monetary profit.

For that reason alone – the contribution that Microsoft and other mega-corporations might make to our efforts in space – attempts to clip their successful wings on Earth are unwise. (The other reasons for leaving Microsoft alone arise from the continued contribution it can make to the information revolution, unfettered; see my remarks in *Digital McLuhan*, 1999.) But such business scenarios for space are nonetheless long shots. In the immediate future – and until space work becomes demonstrably much less expensive – commercial ventures in space may enliven, expand, even goad governmental efforts. But business itself is not likely to shoulder the load.

So, are we back to government – in our case democracy – as the least worst institution to bear the brunt, carry the ball, captain the team of lifting us off into space? Probably so, with apologies again to Winston Churchill, and also for the mix of sports metaphors. But the team metaphor suggests that we can best help ourselves into space not by looking for alternatives to democratic government, but more creatively for its possible partners.

Democracy's partners

So where does that leave us, we who have been waiting a lifetime (and our species a species-time) on the ticket-holders' line to space?

Destructive totalitarian governments are no good to anyone in the long run, neither as standalone mainstays into space nor as partners with democracy; even if they survived, humans would do better to look elsewhere for lifts off of this world. That leaves us with democratic government, spotty as a space mainspring, especially in times of peace. Even in times of war, democracy cannot be counted upon to continue efforts in space – the main responsibility of government in wartime, after all, is to win the war on Earth. Nothing we are likely to discover on Mars or Venus is likely to help the nations of Earth eradicate terrorism. (However, the technology of a missile defense system might be of value in building better space launches.)

So, for better or worse, we have no place to go other than democracy for launching our species more fully into space. Can business be a partner? Yes, as long as we understand that the prime motive for any sustained business involvement, at present, would not be to make a profit.

Are there any other possible players? Any other partners for democracy that can help it focus, help steady its mercurial support as we reach for Mercury, Mars, and beyond? What other institutions, compatible with democratic societies, can marshall the human spirit, train it upon an endeavor that would transform the very nature of our being?

In Chapter 8, we consider the role of religion as a spark-plug for space.

8

OLD-TIME RELIGION AS A NEW WING TO SPACE

On the face of it, here at the beginning of the twenty-first century, traditional religion has little in common with cutting-edge technology. With the exception of the Vatican II Council in 1964 – when the Roman Catholic Church actively endorsed new media as educational tools – organized religion has by and large counseled caution regarding new scientific and technological endeavors, especially those which, like gene-splicing, have the potential for transforming our species. (The strain of fanatical Islam responsible for the September 11 attacks seems amenable to using any technology at hand – but this is in pursuit of political and cultural terror, not religious enlightenment.)

So the attitude of religion to technology has thus far been wary to ambiguous. Could technologies that take us off this planet and out into space evoke a more positive response from Christianity, Islam, Judaism, Buddhism, Hinduism, and like conveyors of thought about the human spirit in the universe? A Web search indicates thousands of comments, papers, and lectures – most by laypeople, a few by local religious officials – that see a place for religion and its values in human settlements in outer space. Could this be indicative of an underlying, widespread spiritual yearning for the universe beyond Earth, in a physical sense, a yearning which could be directed towards helping to expand and sustain our efforts in space?

The thrust of this book is that space as a human endeavor is very different from other technological activities and goals. It is certainly different from all kinds of communication. Thus attitudes of organized religions towards new communications media – ranging from concern about programming deemed immoral, to embracing of broadcast, cable, and Web media as proselytizing vehicles – are probably not very relevant. And unlike artificial intelligence, which can raise questions about human attempts to mimic or create souls, or gene-splicing and cloning, which pitches people in another way into playing roles that some religions might perceive as God-like, movement of people into space poses no threat to religious concerns. Indeed, unlike the nature of intelligent life (and life itself), which science and religions still often dispute – due to the insistence of most religions that intelligence and life partake of a spiritual aspect that science cannot see – outer space has been relatively uncontroversial since the Catholic Church acceded to the cosmology of Copernicus and Galileo. Thus, if movement off Earth seeks to better understand our place in the universe, this is a pursuit of knowledge in no intrinsic conflict with religious teaching and spiritual questing. Indeed, the pursuits coincide.

Moreover, the distinctive transformative qualities of trans-portation – which expand the outlook of the people transported without necessarily changing the people themselves – have made organized religion its champion in the past. Of course, the results of moving people off of this world will be unprecedented and more profound than emigration anywhere on Earth. . . .

Conquistadors to the stars?

The spiritual motivation of the Spanish colonization of the New World is but the most recent example of massive movement of people in at least the partial service of religion. The extraordi-nary initial spread about 800 years earlier of Islam from Africa and southern Europe to the edge of the Pacific would be another.

In both cases, the transportation was intended to convert, and was wrapped in military force. These examples of religious subjugation of new (and, in the case of Islam, already known) worlds are thus akin to the wartime and Cold War motivations of space development in the twentieth century. Terrorism per-petrated by Islamic fundamentalists in the past decade would be another example of religious/military alliances, although the goal is not conversion of non-believers but purging of Western infidel influences from current Islamic societies. Can religion work for space without the point of a sword, gun, or laser?

The bringing of peace to the world was certainly an explicit goal of the Christianity of the conquistadors. That this goal was forgotten or perverted by those conquerors – and at many other times in history – is no reason to conclude that the same would happen were religion to help carry us to the stars. For what awaits us in space is presumably different from what religions encountered here on Earth, in two crucial ways.

First, as far as we know, there are no humans awaiting to be converted in the stars. This would eliminate the main religious motive of the conquistadors in the Americas. Second, there are no humans with rival, militant religions already out in space. This would eliminate much of the motive of the Crusades.

But what about aliens who presumably might have their own views of God, or none at all, and might be deemed ripe for forceable conversion by Earthly religions? Or what about aliens we encounter who might want to convert *us*, at the point of whatever weapons they might have?

And, if we do manage to peaceably settle the stars, what about humans in space in the future, and their possibly militant imposition of their religion on other humans in space?

There are of course no guarantees against any of these scenarios, especially those involving human versus human in space. But there are no aliens, religiously inclined or otherwise, now known to us. And even if we export our religious conflicts to the stars, or sow the seeds of new disruptions out there, we would still be better off than down here.

Coinciding heavens

But lacking any known, possible recipients of proselytization and conversion, why would worlds beyond our planet hold any interest for organized religion? Is the coincidence of interest in the spiritual – in the profoundest of questions, of what we are doing here in this universe – enough?

Religion has actually been interested in outer space all along. It is no coincidence that the Supreme Being, though ubiquitous, is also primarily situated in the heavens above. This signifies a recognition that the deepest aspects of our existence – the source of who we are, and where we are going – reside out there. On that all-important point, the metaphysical heavens of religion and the starry physical heavens of science are the same.

The spiritual importance of space – physical as well as metaphysical – was no doubt behind the Catholic Church's vehement defense of Ptolemaic astronomy against the newcomers Copernicus and Galileo. It was not that the Church had a vested interest in the scientific accuracy of Ptolemy's Alexandrian astronomy per se. It was rather, as Cardinal Bellarmine attempted to argue to Galileo, that the Church saw the stationary position of the Earth at the center of the Universe as necessary for the spiritual (and what we would today call psychological) health of humanity. Bellarmine struggled to find a way to reconcile Galileo's telescopic observations – foolishly attacked by less learned men of the Church – with those spiritual needs. Bellarmine failed, because physical reality was too fundamental to spirituality, in the Church's view, to be really negotiated.

Indeed, the literal, physical return of Christ after the Crucifixion – Christ returned not only in spirit, but flesh – is crucial to Christianity, as Elaine Pagels observed in *The Gnostic*

Gospels in 1979. This gave physical reality a place far more important in Roman Catholicism than in religions such as Buddhism, for example, and may even help account for the empirically scientific and technological revolution in the West (see my "Easter Theory of Technology," in *Electronic Chronicles*, 1992), even though that revolution was initially at odds with the Church, and the rift continues in such areas as cloning.

So religion and science went their separate ways – likely to the detriment of both – with Galileo and Bellarmine unable to agree on a physical universe that would accommodate the Church's view of spirit, and with Darwin widening the schism for different reasons two-and-a-half centuries later.

But the crucial point to record about those developments here is that the schism was based on a shared passion about the importance of the human place in the cosmos.

Indeed, religion and science can cooperate today, give expression to their mutual passion for space, in ways they could not 400 years ago. The conversation between Bellarmine and Galileo was unequal, since the Church at that time still had enough temporal power to threaten Galileo's work. Today, the Church holds no such power. No religion does, except perhaps for the fundamentalist dictatorships in the Middle East over their peoples. This means that religion and science can be aboard the same spaceship, equal passengers with different proximate purposes, in pursuit of the same ultimate cosmic questions.

How would religion make its contribution?

Socially speaking, the military can get things done more easily than most other institutions: orders are given, and the chain of command responds. Democracies are messier – uncertainty of outcome, unreliability in follow-through, is the price paid for safeguarding of freedoms to disagree in public discussions designed to elicit consensus or agreement. Businesses operate on a Darwinian level when the government allows: in the commercial marketplace, the strong and attractive survive.

Religious institutions are like none of the above. Although the faithful might seem to respond in a lockstep military manner at times, the truth is that the faithful have to be inspired if they are to perform continually. Unlike the professional soldier, the religious faithful have to believe.

How would religions encourage their followers to believe in the need, the sanctity of space exploration and settlement? The process would no doubt be gradual and long. It might start with the Pope talking about how movement into space was spiritually appropriate, and consonant with the Church's mission of uplifting the human condition (in this case, literally and figuratively). It could begin with a council of rabbis pointing out that human improvement of the universe – leaving it a better place than one found it – is called for in the deepest Jewish teachings, and human presence in the universe beyond Earth might be a good addition to what we already try to do on this world. Voices could be raised in Quaker gatherings; Islamic mullahs could speak of outer space as Allah's vistas; the Dalai Lama could publicly muse about physical ways of achieving transcendence off of this planet. Most of this is of course unlikely to happen. But if even just some of it did, a new element would be injected into the equations for space – an element which, by its very nature, would not be bound by practical considerations, not measured by immediate military, scientific, or commercial dividends. Religion would indeed introduce an element that was closest in its form and reach to unbridled human imagination.

The role of imagination, of wish-fulfillment, of wanting dreams to come true, is often misunderstood in human affairs. As vehicles for getting things done, wishes cannot usually accomplish much on their own. But in concert with science and other practical partners, they can and do move mountains. Every technology ever invented is the embodiment of a wish to see something done in a better way – along with an embodiment of further ideas about how that wish can be implemented (see my *Mind at Large* (1988) for development of this view of technology). On this reading, we might say that one of the problems of cyberspace is that it is just another form of unimplemented, unembodied wishing – in contrast to realspace, a place where wishes are embodied. To get into an automobile is to enter a compilation of wishes not only about transport (combustible engines, wheels, and so on) but comfort (seating upholstery, air-conditioning), scenery (windows), news and music (car radio), and so on – wishes that have all come true via their physical embodiment in technology.

Sometimes this technological seriousness of wish-fulfillment is not understood; imagination is seen instead as frivolous, childish, unrealistic. Sometimes even champions of science fiction have been given to say that it must decide if it will deal seriously with issues of humanity in the cosmos, or cater instead to adolescent wish-fulfillment. But this misses the point that wishing can itself be the cosmos calling for attention, as it expresses through our intellect its/our need to be in physical proximity to more of its span than available here on Earth. And if this seems juvenile or childish, that is because our adult pursuits, our daily, proximate, important business, all too often leave little room for such callings.

Religion, whatever its other strengths and weaknesses, encourages such stirrings. The wish-fulfillment in religion may be difficult to notice – hard to equate with the imagination of science fiction – because the wishes in religion are so ancient, so deep, so ubiquitous. But they are there. Religion may be seen as an institution for the constructive channeling of such wishes. What I am proposing here is that the wish to get off this planet, and out into space, be adopted by religious institutions – because, in fact, this wish is already implicitly a part of most religions.

Embodied in technology, driven by quest for knowledge (science and philosophy) and spirit (religion), with pursuit of a little eventual profit (business) and the physical highground (military) – always valuable for defense of democratic societies given the enemies of freedom that our world all too often spawns – our wishes for the stars receive their best direction.

But all of this profundity of wishing for space does not mean that space, and wishing to get there, cannot also be fun. And while it is true that imagination is by no means only for children, it is also the case that some of the qualities of childhood pertain very much to human entry into the cosmos beyond Earth, and recognition of such qualities may help us get there.

Playful space

The religious view that we are all children of the Universe coincides with wishing upon a star, and the joys of imagination, as wellsprings for outer space. Part of being a child entails a playful openness to new possibilities: indeed, also to near and outright impossibilities. But the child cannot usually do much about imagined possibilities. And adults, who may have the wherewithal to break out of the mold, often lack the unfettered imagination. Sustained movement into space will likely require both.

Freud captured part of this tension between the creative, powerless child and the uncreative, powerful adult when he noted over various essays the similarity in the psychologies of children, neurotics, "primitives" (people living in the least "civilized" states, for example, the Australian aborigines) and artists. All four groups have high degrees of unimplementable imagination. The child and the "primitive" both believe in magic – the child, in part, because he or she lacks access to adult modes of accomplishment, the primitive because he or she lacks access to scientific and technological modes, or the same modes that are beyond the control of children in more civilized societies. The neurotic adult could have such access, but problems of social integration and communication get in the way.

That leaves the artist. Here, we might make a distinction between artists who starve and those who succeed. The first indeed bears resemblances in Freud's perspective to the child, the neurotic, and the primitive: all four partake of unrequited dreaming. But the second kind of artist – who succeeds – is something else. For the accomplished or acclaimed artist can be seen as an adult with a child's imagination and an adult's effectiveness – in this case, the talent to express one's imagination in an art appreciated by other adults.

This is precisely the kind of fruitful, rewarded dreaming we want for space. For humanity to live beyond this planet, our cutting edge – our leading dreamers – will need to be the equivalent of Shakespeare, Picasso, and the Beatles, all artists who have made their dreams come true. Of course, this was not easy, and their success required not only talented dreams but perseverance and luck.

But art, in its playfulness, can have an additional role in helping us into space. Art can inspire those of us who are not artists about the excitement of the Universe. In this role, art has it easier than religion. Everyone is a believer when it comes to art. Although different artists and genres of course have their followings, all human beings respond to one kind of art or another. (I suppose there could be the equivalent of an artistic atheist or agnostic – someone who denies or doubts the persuasions of all music, paintings, novels, motion pictures, and so on for him or herself and others – but that person would be far rarer than the doubter of religion in today's world.)

What sort of art can best lure us off this planet?

We have already seen a little of what popular culture – motion pictures and television – can do. *Star Trek*, *Star Wars*, and their progeny, noted earlier in this book, have no doubt stirred more interest in space, more comfort in our eventually being there, than any real expeditions in space since Apollo, and indeed perhaps more than was engendered by the Moon walks themselves. The reason is that the dramatic presentations are imbued with human elements – romance, suspense, humor, friendship – which real excursions into space usually cannot present as effectively. An advantage of art over reality, of painting over photography, of novels over journalism, has always been that art can make the reality more humanly appealing, engaging, gripping. This pertains even to life-and-death situations, which are always the backdrop of real journeys into space, but are usually portrayed more plangently in the fiction of pages and screens. The exception would be when catastrophe strikes a real excursion, and lifts the life-and-death aspect of all space work from the general background to the harrowing foreground. Otherwise, entertainment about space usually wins over news.

In Chapter 9, we consider what may be done to make our everyday, real encounters in outer space more compelling.

9

WOULD YOU WANT TO LIVE NEAR A STAR NAMED HD 209458?

Semanticists – those who study the psychology of naming – like to distinguish between the map and the territory, by which they mean the label and the reality it describes. In general, we want our labels to be as accurate as possible. And we do not want to lose sight of the difference between labels and realities. Cursing someone – an act of insult through words or labels – may be reprehensible, but it is not the same as physical assault.

Nonetheless, as this very example suggests, labels may well lead to actions, for better or for worse. Fighting words may trigger physical assaults.

What labels are most likely to coax us off this planet and out into space? As of December 2001, astronomers have spotted some sixty-six planetary systems, sporting seventy-four planets (seven of the sixty-six have more than one planet). Most of these suns have designations such as HD 209458. A few are more interestingly named: "Tau Boo" (short for "Tau Bootis") is about fifty light years from us, and is thought to have one planet (currently designated "Tau Boo C"), more than three times the size of Jupiter, circling around it at 0.0462 AU (distance of the Earth from the Sun). "Ups And" (short for "Upsilon Andromeda") is about eighty-two light years from Earth, and is thought to host three big planets (the smallest, about three-quarters the size of Jupiter, circles at 0.059 AU).

Those are about as good as it currently gets – namewise – for new places in space.

Are these names scintillating, suggestive of homes away from home, places far from our world where we might feel comfortable some day? Hardly. They are better than lifeless astronomical labels (HD 209458), certainly. But beyond that? "Tau Boo" and "Ups And" are hold-overs from Greco-Roman names for constellations, and are intriguing only insofar as the nicknames ("Boo," "Ups And") strike some chords in current English usage. "Boo" and "Ups" are miles above "HD 209458," which has all the appeal of "CP/M" and "MS-DOS," names of early personal computer systems. But "Boo" and "Ups" lack the straightforward, instantly recognizable invitation of "Windows."

Of course, Jupiter-size planets would probably not be readily inhabitable whatever their names. So maybe "Boo" is aptly named, signaling its single, gargantuan Halloween planet, of which we should be wary, if not scared, at least as a place to live. But surely we can do better for Earth-size worlds, revolving at distances appropriate for Earth life around their suns?

In fact, the names of constellations, in ancient through medieval times, had the effect of humanizing the heavens, of endowing them with human attributes, actions, and relationships. They seem out of touch now, in part because most of the Greek and Latin have no immediate meaning in our world (for example, "Epsilon Eridani"), and in part because the constellations themselves derive from observations of stars with our naked eyes, and not where our powerful telescopes show the stars to actually be (stars in the Andromeda constellation range from 72 to 390 light years from Earth).

The names for planets in our own solar system also come from Greece and Rome, but are so well known that they are in effect part of English. And, too, they have the cachet of vividly human, heroic characteristics – speed for Mercury, beauty for Venus, and so on. But the names of the best-known Greco-Roman gods are already taken. And with numbers of newly identified planets increasing by as much as four per month (the total jumped from sixty-six to seventy-four between the end of September and the end of November 2001), we would quickly run out of names even if we resorted to lesser-known gods, and even descended to sprites, nymphs, and demi-gods.

So what can we use as a source of names for the stars? What dictionary can we consult for proper cosmic titles?

Once again, the New World on Earth – the Western Hemisphere – may help show the way.

Old names for new worlds

The names of places near to my home – New York, New Jersey, New England – suggest an atlas of appellations for the stars and the planets that may revolve around them. There is nothing particularly poetic or original about such names. To the contrary, they invoke the old and the comfortable. Yet therein resides their power. Faced with huge tracts of unknown, potentially hostile territory, the Europeans who settled in the New World often chose names that recalled homes left behind on the other side of the Atlantic.

Of course, some Earthly names would work better than others in such a schema – "New Chicago" would certainly scan better than "New New York" as the name for a habitable planet (and if this planet was prone to gusty winds, so much the better). But "New Manhattan," "New Brooklyn," and "New Bronx" would still be contenders, or we could call a new world "Newer York" and still get all the benefit of that New York moment. (*Newer York* was actually the title of an anthology of science fiction stories about future New York, edited by Lawrence Watt-Evans in 1991.) And should we run out of local places, we can turn to names of people – astronauts, cosmonauts, scientists, political leaders – who fostered space exploration. This approach is already in evidence on the Moon, where Copernicus and Tycho live down the street from Mare Tranquillitatis.

If all of this sounds suspiciously like an ad campaign, an attempt to come up with an appealing name for a new product so that it will sell, that's exactly right. The worlds in space are indeed new – at least, to us – and they need to be sold like toothpaste. Millions of dollars are lavished on the creation of names for everything from corporations to cosmetics. Why not devote at least a little of this attention to names for space, where the stakes and the opportunities are so much greater? In other words, the fact that toothpaste and cosmos are at opposite ends of the spectrum of profundity only makes it more imperative that we do everything in our power to sell the cosmos – meaning, make it attractive as a place to inhabit – including picking up whatever pointers we can from toothpaste.

Advertising campaigns for new products, just as names for new places, employ the strategy of metaphor, as ancient as human language itself. Derived from the Greek word "metapherein" – to transfer, to bridge, to move across – metaphors help us come to terms with the new or unusual or even unfathomable by connecting it to aspects of existence we already understand or have at hand. We grasp something of the swift, sometimes inexplicable passage of time when we say "time flies" or "time's running out." In reality, of course, time neither actually flies nor runs – birds and animals and insects do. But by investing time with these common properties of life, we can better understand, or at least discuss, the effect of time upon us – and thus how we can put time to our best use.

In the case of space, metaphors can serve this role not only in making us more comfortable with strange places, but also with the process of getting there. It is just a question of choosing the best metaphors – the right familiar things to which the movement of humans in space can be connected.

Spacefaring metaphors

Science is a mode of discovery; military action a way of conquering and defending against adversity. Neither are alien to humans in space. Indeed, our enterprise in space may be aptly described as discovery of the little or not known, and the mastery of those parts of the poorly known that might be dangerous. (Military action is usually directed against dangerous humans on Earth, but there may well be dangers that need to be defended against, even sought out and destroyed, in the vast reaches of space.) And yet science is not the only way of increasing and assessing our knowledge, and adversity may be encountered and engaged in ways other than military. Thus the need to go beyond "surveying" and "colonizing" when we speak – and think – of humans in space.

The old-fashioned notion of "space travel" actually works very well, because it says in plain, homespun, readily comprehensible terms what we are doing in space. Every astronaut who flies off of our world in a shuttle is a space traveler. So were the people who went to the Moon. So will be the civilian scientists and military officers who travel to Mars – space travelers, all of them. But focus on the scientific expertise and military prowess of people on these expeditions, certainly not incorrect as descriptions of who they are, can distract from their more fundamental role as space travelers.

Even the term "expedition" may not be the best, though it is sometimes preferable to "mission" and its military connotations. When we travel to and fro on our world we do not usually think of ourselves as taking or conducting an expedition – that is a term usually reserved for climbs of Mt. Everest, or journeys to remote parts of the planet. But isn't a journey anywhere in space far more extraordinary than a climb anywhere on Earth? Of course. The label "expedition" – and even "mission" – is not inappropriate for journeys through space. What is instead being suggested here is that they may not be as effective as a mundane word such as "travel" for encouraging our species to think of human life as naturally, appropriately — eventually normally — extending beyond this world.

But why would any mundane term be useful, even necessary, in our thinking and discussion of space, given that the very word "mundane" means "of this world"? The answer, if we consider "space travel" as an example, is that the word "space" already conveys more than sufficient wonder, excitement, grandeur. Space "expeditions" thus add no extraordinary quality or implication not already present in "space." But "travel" contributes a comforting quality missing in "expedition." We can not only afford this mundane balance — this sense that we can eventually do in space what we do on Earth, just on a more majestic scale — but we may need it, if space travel is ever to move beyond science fiction, and scientific and military operations, to a way of life.

And there are things we can do to make the universe at large seem more like home.

The comforts of home in space

On the one hand, the Universe beyond our planet seems the epitome of inhospitable: certainly nothing is breathable in the vacuum of space, and we have no reason to think anything is breathable — not to mention directly potable or edible — on worlds and moons beyond our planet. Certainly our own

Moon, and other worlds in our immediate solar system, have foreboding environments, inhospitable except through the utmost of current technology. Yet we here on Earth are nonetheless residents of this greater universe – the Universe is our home. So how do we go about making it really *feel* more like home – not just in terms of its packaging, as discussed above, but in reality?

We faced an analogous, if less severe problem on our own world. Although there is air just about everywhere above the sea – it gets a little thin at very high altitudes – the same is not the case for food and water. And temperatures are forbiddingly cold for part of the year everywhere except the tropics.

Technology of course made the difference here on Earth. As a strategy of life, technology – the rearrangement of material to serve organic ends – considerably pre-dates humans in the evolutionary schema. Not only do chimpanzees use sticks to ferret out termites, and beavers build dams, and birds make nests, and spiders secrete webs, but even viruses that move one bit of some organism's DNA from here to there, with whatever purpose or effect, are engaging in a technological enterprise. Chimps, beavers, birds, and viruses are creatures of technology, one and all (or, in the case of the virus, something less than a creature). Technology has antiquity, not only in human but in biological history.

But we humans do take this truly ancient enterprise of technology to a higher plane. Our use of technology remakes the world. We can change the forms of life itself by the invention of genetic engineering – updating and expanding and focusing, in effect, what the virus has already been doing, lo all these many millennia. We can also now literally leave this planet via technology, a development that goes beyond the flight of bees and birds that inspired our first artificial excursions off the ground, via glider wings, balloons, and airplanes.

The difference between human and all other organic technology is that ours embodies our intellect, rationality, imagination, capacity for dreaming, loving, and (in the case of weapons, or technology used as weapons) hating. The same technology can often be used for opposite purposes – loving and hating – whatever initial ideas it embodies. Thus a knife may be used to cut food for a child, and a knife may be used to stab someone; a plane may be used to facilitate a visit to a loved one, and a plane may be used to kill thousands in a skyscraper. Communications media are especially prone to this Jekyll-and-Hyde usage – we can use the phone to say "I love you" or plan a heinous crime, just as we can through e-mail or just plain in-person conversation. But all other technologies are also knives that cut both ways, including those that at first seem intrinsically good or bad. Thus medical technology, working to destroy germs, becomes germ warfare; and the science of the atom bomb engenders radiation treatments for cancer.

With this in mind – the accidental, unintended consequences of technology, mixed with our capacity to apply technology to whatever very specific purposes we desire – we can better fashion and enlist technology to the task of making the stars our home. We can plan to make the Universe, or at least given places – other planets – in the Universe more like Earth.

Aside from the technological prowess needed for such terraforming – most of which we do not yet possess – such planning first requires our thinking and speaking of outer space as home. This begins with a recognition of the transformative powers of technology – a recognition that the very human business of technology is turning dreams into reality – and proceeds to a selection and nurturing of appropriate dreams. We can do miracles with technology, and the miracle we want in this case is the extension of human life into space. And we want to enact that miracle in a way that makes it least likely to be used for some destructive purpose – foreseen or otherwise – although we also know, realistically, that there are no guarantees regarding application of technology.

This is why language and metaphor are so important. We are not likely to build a bridge to anywhere unless the gap is first spanned by metaphor. We are already in space on scientific missions, and these are enormously valuable. We have also been in space, and will no doubt continue to be, for military reasons. But we are less likely to ever live there, beyond our planet, if such residence is left as an unintended consequence of our science and military defense. Instead, we need to pursue science and defense in space in the larger context of providing knowledge not only of the stars, not only for what they can tell us of who we are, not only to help us get there, but to live there – assuming, of course, that this is what at least some of us want.

But if proper language is necessary for the humanization of space via technology, is there anything in our specific choice of technology that can enhance this pursuit? Since we are still a fair way from actual terraforming, this question of best devices amounts to what are our best scouts, the best vehicles for our first encounters out in space?

In Chapter 10, we consider the proper role of robots in ventures beyond our planet.

10

REAL ROBOTS DON'T CRY

Actually being there – physical transport and presence – is ultimately a more reliable path to knowledge than what can be conveyed through just images, sounds, and data. Even the most multi-directional microphones and cameras can be deaf and blind to something subtly significant, right in front of them, that would catch the attention of a human being on the scene.

But the problem is that getting a human being on the scene is no easy matter when the scene is off planet. For some places – indeed, anywhere beyond our solar system and its suburbs – getting a human being on the scene is currently impossible, because, for starters, we lack the transport technology to get any physical object out there. As an example: Pioneer 10, launched from the Earth in 1972, traveled more than 7.5 billion miles by 2002 (Pluto is about 2.7 billion miles from Earth); 7.5 billion miles is about eleven light hours from Earth (the time it takes light, traveling at 186,000 miles per second, to cross that distance); Alpha Centauri, the nearest star to Sol, is about 4.3 light years away. In the case of Alpha Centauri and every star other than ours, far-seeing telescope technology and the data it delivers are thus all we have. Physical transport to those places awaits radically faster means of propulsion.

For some destinations closer to home, in or near our solar system, we may have means of transportation, but it is not suitable for transport of humans. We can send a device to Saturn, for example, but not a ship – at present – that can sustain and return human life. This is in part because of the years it would take a vehicle powered with current technology to reach Saturn, and in part because we have not yet sent any object to Saturn and brought the object back home. Indeed, thus far, the only extra-terrestrial body to host human visitors who have returned to Earth has been the Moon. Mars seems the next likely host.

In this chapter, we consider the service that robots, as we currently know them, can best provide for these encounters in different parts of space. Our answers will hinge, in part, on whether today's and tomorrow's robots are more like people or cameras for the purposes of space scouting and exploration. And our considerations will not only be practical (to what extent can robots fill in for humans – perform in the same ways) but ethical, since some people (for example, Alex Roland) have argued that robots are preferable to humans not only in deep-space but even in all near-space encounters, in part because robots are more expendable than humans.

These things we already know about robots. They move

through the environment, and on that score are much more like people than telescopes, cameras, microphones, and similar data-collectors, either stationary, or mobile only if borne by a human being. So robots are preferable to all media of observation (telescopes), communication (cameras and microphones), and analysis (computers), except when such media are either too big to be incorporated into a mobile robot capable of being transported on a spaceship, or the celestial body under scrutiny is too far away to be reached by any known mode of transport. Examples of such intrinsically immobile media, at least by current standards, would be huge telescopes and computers. Certainly they exceed the carrying or incorporating capacity of any robot the size of a human, or even an elephant. However, the Hubble telescope suggests that, somewhere along the line, big telescopes might be able to function as spaceships. When that happens, we may well have a combination robot plus telescope plus computer plus spaceship, in which the big-media carrying problems are solved.

Until then, we can at least say that robots are preferable to all smaller, stationary media for places in space that are within travel range.

But are robots preferable to people?

We begin with robotic advantages.

Robotic merits

The first and foremost advantage of robots over humans as explorers of space is that the loss of a robot is at worst a severe inconvenience of time and money – it is not a loss of life. Here on Earth, in environments in which we already know, and know to be dangerous, use of robots certainly makes sense. If we could construct a robot to mix dangerous chemicals, or biological materials, what logic would there be in insisting that human beings do the job?

There are also some lesser benefits of robot workers, which might well commend them as explorers. Robots don't get bored with repetitive tasks. They don't miss family back home, don't fall in love with co-workers, don't get angry, annoyed, distracted with any of the myriad concerns, small and monumental, petty and wonderful, to which we humans are prone. In short, robots partake of all the advantages of strictly logical, unemotional artificial intelligences – computers – that operate wholly on the basis of programs and circuits not hormones.

The combination of such advantages – real, if difficult to calculate – along with the cardinal benefit of robots not having a life to lose, make an undeniable case for major robotic participation in our engagement of the solar system and its worlds and moons. But should this participation be to the exclusion of flesh-and-blood humans?

The answer would certainly be yes, as long as there were no significant advantages of humans over robots in space. The fact that a human has a life to lose, and a robot does not, is clearly a weighty argument against sending a human to do a robot's possible job.

But does being alive – being fully human – also have some irreducible advantages for the job of emissary to the stars?

Limitations of programming

Behavioral psychologists such as B.F. Skinner once argued that humans are no less programmed than machines – any difference resides in our unawareness of our programming (that is, we are aware that we have programmed the machine, but are unaware that we have been programmed by our past experience; the unawareness of the machine or computer of its programming by us is not at issue). Marvin Minsky made a similar point when he referred to our brains as "meat machines" – by which he meant the significant difference between our brains and computers is that we are constructed of protein and computer chips of silicon, but both run on programs. (A similar perspective animates the distinction between cyberspace and "meat-space." The latter refers to what I have been calling "real-space" in this book, but seeks to deprive it of any privileged position in human affairs over cyberspace. I prefer my appellation "realspace" precisely because it calls attention to that position.)

Characterizations such as Skinner's and Minsky's are often presented as alternatives to the religious notion that we – our brains and bodies – are somehow charged by an ineffable soul, or the philosophic contention (from Descartes and many others) that our brains are driven by, and/or give rise to, our nonmaterial minds. In the homogenized programmed mindset, thought and dreams and schemes and if/then algorithm routines are indistinguishable in their essential nature.

But the claim that human mentalities are totally the product of programming, in the same sense as a computer's (or robot's) decisions, can be refuted with recourse to neither soul nor mind. The plain fact is that the human adult has had experiences which verge on the infinite in comparison to anything programmable into a computer, and these human experiences play upon genetic "programs" in the brain – such as our capacity to communicate via language – that are themselves vast and barely charted at present. Thus, even if we want to accept, for the sake of discussion, that human mental operations are "programmed," there is no meaningful equation or even comparison between that kind of human programming, and what we do with computers. The scales of relevant programmable variables are in two different universes.

For all intents and purposes, then, human beings are in effect "unprogrammed" in comparison to computers. Imagine a person standing on a new world, or looking at it from a hovercraft, for the first time – or, for that matter, a second, third, hundredth, or thousandth time. Think about what this visitor might see after a moment, an hour, a decade, a lifetime of observation on this new planet. Perhaps some nuance of movement, some flicker of pattern, that might resonate with something she saw when she was 5 years old, walking to school, playing in a garden, climbing a fence, back on Earth. Maybe a ripple in the new world's water, a bend of a leaf, that recalls something she saw when running with her friends by a lake in the park when she was 11. Or perhaps the glint of the sun in this distant solar system, bouncing off a mountain for an instant in the morning, that conveys something of her grandmother's smile.

Does this last example, especially, take an unfair shot at computers? Does it have little discernible connection to scientific inquiry? That is exactly the point. Scouts need more than rigorous scientific education. They also need wellsprings of unspecified emotion and experience, the better to match up with who knows what may be encountered out there in space. What computer program embedded in a robot's brain could possibly encompass such experience?

This near infinity of uncataloged experience that all humans carry is a friend to discovery. The key difference between humans and robots on this point is not that our minds are more specifically prepared than robot programs to recognize important details and events. Rather, our advantage comes in what we have not been prepared for – in the plethora of experience of just living and thinking. If Plato is right in his "Meno" paradox that we cannot learn anything new unless we already know something of it – for how else would we recognize the new as knowledge at all? – then the grab-bag of life is surely indispensable packing for voyages of discovery in space.

And we humans have other advantages over robots.

Knowledge in tiers and tears

The three traditional branches of philosophy ask what is true and false (epistemology), right and wrong (ethics), beautiful and ugly (aesthetics). Computers may be programmed to make decisions about all three, but are on weakest grounds in the third area. Facts of course can be stored in digital, non-organic forms, and accessed by programs which can then render informed decisions about whether new data are true or false; this can certainly be of assistance in epistemology. Ethics are more difficult, but amenable, to some extent, to if/then programs. A computer "judge" could consider: If a man's 3-year-old daughter is in need of urgent medical attention, should the man be permitted to exceed the speed limit to rush her to the hospital? Suitably stocked with if/then guidelines and examples – if medical needs are sufficiently urgent, then traffic rules can be violated – we could have confidence that our AI magistrate would make the right decision.

But aesthetics? A robot might well be programmed with an immense store of images and sounds that humans have judged to be beautiful. But would we then trust it to recognize even a slightly new form of beauty?

The problem is that the principles of aesthetics are poorly understood even for humans – sorting through facts and weighing of ethics primarily entail thinking (though some feeling or emotion may be involved in ethics), but objects of beauty are felt (apprehended by emotions, as some old-time philosopher might have said), with perhaps some thought involved. So beauty is quite the opposite from truth and right in the blend of intellect and emotion it requires to be perceived and assessed.

Consider the plight of the robot looking out at some desolate landscape in some faraway part of the galaxy. The very word "desolate" underlines the robot's problem. Presumably the robot could give us a factually accurate report of the paucity of stars, planets, mountains, whatever, in the area – how many it observed, and how closely packed. Our robot might also make an ethically sound decision. Say, if a human explorer died, and uttered nothing before dying, our robot might be programmed to nonetheless tell the explorer's husband back on Earth that she said she loved him just before she died. But what could our robot report – what could it know – about the aesthetic state of "desolate"?

Could it convey anything of Pascal's sense of the awesome disproportion, the minuscularity, of our minds, brains, dreams, our existence, in comparison to the immensity of the Universe? Could it know the awe that we feel right here on Earth when we look out at the starry skies? Not likely. No, not at all. Our robot might well know of *our* awe, and the circumstances that call it forth. Our robot might notice something different in our eyes when we look up at the Universe from this Earth and try to fathom what it all means. Our robot might well be programmed to recognize similar circumstances in other places – jet-black nights agleam with silent stars over deserts or tundra on other planets – and tell us about them. But to know the awe itself, to know the sense of desolation, would require feeling, experiencing the awe – and that, in turn, requires emotions, or brains bartended by adrenaline and other hormones. Can logic circuits be bathed by hormones – meaningfully "interface" with them (to use that awkward verb)? Perhaps, someday. But certainly we have no robots or computer programs which feel that way today.

And the presence of human emotions on the cutting edges of our expeditions in space is no mere luxury. How many times have you walked into a house or an apartment you were thinking of buying or renting, highly recommended by a friend or agent and/or looking great on paper, and, as soon as you were inside, you realized that it was not right for you? How many times have you had the same experience with an automobile you were considering purchasing? These decisions are based on more than knowledge, facts, information. They entail emotional responses, sometimes so minute as to be unconscious, or incapable of articulation. And yet – certainly in the case of a new home – they can guide the most important decisions of our lives.

Don't we need every bit of this as we consider new homes in space?

Robots with emotions?

But if we are thinking about distant futures, might we not be able at that time to build robots with emotions? It isn't so hard to imagine – indeed, it's been imagined many times in science fiction literature, and well portrayed on the screen.

If we were able to someday construct robots to not only reason, but feel, would these robots be subject to the entire range of human emotions? Anything less would still count against them as our sole envoys to the stars, for we would be hard-pressed to know just which of our emotions were most needed for interstellar excursions. And creation of robots with full ranges of human emotions would presumably be more daunting than building of robots with only some emotions. The portrayal of robots and androids (robots that look like humans) in motion pictures and on television actually makes such a task – the creation of fully "feeling" robots, with computer programs capable of complete aesthetic responses – appear far more easy to accomplish than it probably will be. Human actors and actresses are, after all, portraying the androids, so any emotion we may see in the androids is a function not of the android (which of course does not really exist as yet) but the face of the human actor or actress (which of course really does exist).

Specialists in nonverbal communication have cataloged hundreds of different smiles, smile/eye combinations, and so forth on the human face. Most of us not only make, but recognize and understand, most of those intricate, subtle expressions. A robot could probably be programmed in the recognition; even the rendering of the expressions seems not beyond sufficiently sophisticated android construction. The catch is that each expression, for us, reflects an internal mental state – an aesthetic or other emotional response. What would the expressions reflect in a robot?

But let us assume, for argument's sake, that we will someday be able to construct robots with the entire panoply of human emotions. Would such entities then be preferable to humans as our sole emissaries to the stars – that is, could we be comfortable with their observations and reports and on-site analyses and decisions, in our human absence? Equating emotions with souls for a moment (or souls with a mixture of cognition and profoundest feeling which characterize our humanity), and putting aside the religious question of how souls come to be in the first place (for we could say that, if only the Supreme Being can breathe souls into humans, then these robots were so human in every other way that the Supreme Being might have seen fit to invest them with souls), would robots with such "souls" make satisfactory sole envoys – soul envoys – to the Universe?

Sadly, no. And not because such robots, on this account, would still be less than human. The problem with robots "who" were human in every important way and went to the stars on our behalves arises from just the opposite – the fact that such hypothetical robots would not be different enough from us. The one undeniable advantage of any robot, as we saw above, is that we can feel comfortable risking its existence before that of a human being. No one likes to lose a car or an expensive computer, but only a lunatic would hold such losses equal to that of a human being. So, if we make robots so human that they can survey the Universe in our complete absence, without any detriment to the results of the survey, are we not therein making their loss – their "death," perhaps – no less unacceptable than ours?

We may think of this as a fundamental, unavoidable, paradoxical "principle" of robotic substitution for human beings: the more like humans they become, the less acceptable their exposure to danger, injury, death; and, the less like humans they remain, the less reliable their reports about places we have never seen or experienced first-hand. This limitation gets us and our robots coming and going. It counts against their use as our sole emissaries at all stages of their development, primitive and present and real, as well as future.

Science fiction, for the most part, has focused on the dangers robots may pose to us, not what perils we might visit on them. Isaac Asimov's "three laws of robotics" – a robot may never through action or inaction cause harm, or permit it to come, to a human; a robot must follow all human orders, except when they contradict the first law; a robot must act in its own self-defense, except when such actions might contradict the first two laws – are an excellent example: a robot's well-being comes last, even subsidiary to a moronic or malicious human order to jump off the roof, for no reason other than that the human has commanded this. (Asimov later posited a "zeroth" law, to take precedence over the first, which insisted that the good of humanity as a whole outweighed the needs of any human individual. This could allow a robot to kill a person on behalf of humanity. But the new law was no help to robots, whose "rights" were still in last place.)

But the more intelligent, profound and sensitive we build our robots – the better able they may be to implement Asimov's laws – the more in need of protection they themselves become from us. Because they have become, or are becoming, us.

All things considered, then, we may have to settle for low-grade robots – low-grade in terms of science fiction, high-tech in terms of what we may actually now or soon produce – to accompany us to the stars. They present us with no ethical dilemmas in their loss as they help us with whatever tasks we assign to them. The only danger we court in their use is assuming they can make decisions, fully explore worlds, in our absence.

A significant advantage of looking at robots in this way is that they are already to some degree attainable, or nearly attainable, with our current or foreseeable technology. As much as we need visions of the distant future to guide us into space, we are unlikely to get very far unless our plans are shaped in terms of technologies at hand, or nearly so. A serious reorientation of our species towards the stars requires taking account of where we are right now.

And thus, in Chapter 11, the concluding chapter of *Realspace*, we consider how the events of September 11, 2001 and their aftermath have made the arguments in this book even more pressing.

11

REALSPACE IN AN AGE OF TERRORISM

The events of September 11, 2001 instantly changed the beginning of the twenty-first century from the Age of the Internet to the Age of Terrorism. Or, on that day, the Internet's claim to the spirit of our age was disputed. The nature of life on this planet was altered.

Ironically, and importantly, the Internet itself was not the target. With all the buzz about the "millennium bug," all the concern about possible plans of demented hackers to bring the world economy down, and perhaps the world itself, by attacking our information systems – the nervous system of our planet – the assaults of September 11 on the World Trade Center towers and the Pentagon were physical. Planes, vehicles of transportation in physical space, crashed into buildings and killed thousands of people.

This represents the ugly aspect – the ugliest aspect – of

realspace. Planes exploded, buildings crumbled, people died. No event in cyberspace or virtual reality – the magical kingdoms of indirect, second-hand reality – could have had such a horrible, tangible, down-to-Earth effect. Among the bitter lessons of this tragic trumping of communication by transportation – twisted transportation – is that we have to pay more attention to the real. Information has its ultimate impact, as we have discussed throughout this book, in the world of people and physical objects in realspace.

In this final chapter, then, we consider what the world that September 11 brought into being can teach us about communication and transportation, about the Internet and the cellphone, and about movement of people around and off of this planet. And we consider what those lessons can in turn teach us about our perplexing new world.

Two-edged swords of communication

We should not be the least bit surprised that terrorists use any communication means at hand – the Internet, satellite cellphones, videotape, scribbles on a piece of paper, just plain talking. It was always so, and always will be. Any device that facilitates communication can make murder, mass murder, any heinous crime that requires planning, a lot easier.

This seems to have been better understood about telephones than about the Internet. Prior to September 11, the American government's prime attempt at policing the Internet was directed against pornography. Unlike wiretaps on telephones, which at least attempted to combat drug traffic, organized crime, and terrorism (alas, the first two more successfully than the third), the Communications Decency Act of 1996 went after websites with dirty pictures. It was wisely struck down as unconstitutional – in violation of the First Amendment – by the U.S. Supreme Court. But the response of Congress was only to try to devise new legislation that would regulate porn on the Web and pass Supreme Court muster.

Terrorism, of course, has no doubt been around for as long as pornography, which apparently goes back at least as far as the ancient tombs of the Pharaohs. Certainly terrorism beset the twentieth century. Yet no one in law enforcement mounted a major campaign to interdict terrorist communication on the Internet. Pornography, and, to a lesser extent, business fraud, received the attention and, where possible, the arrests.

September 11 changed that focus – too late for its thousands of victims, but not, one hopes, for the future. Eavesdropping on the Internet – government interception and reading of e-mail, via the FBI's "carnivore" program, which keeps the sender and receiver in the dark about the interception – does raise new freedom of speech issues. But no communication system was ever safeguarded to serve as a vehicle for hatching and coordination of conspiracies to commit murder. Unlike pornography there is no ambiguity, no rights of expression that may need protection, in the planning and commission of mass murder.

On the positive side of the digital ledger, the Internet's decentralization, among its most significant characteristics, enabled it to serve as a backbone of national and international communication for many whose phone service was impaired in the hours after the WTC attacks. The genius of "packet switching" on the Internet is that any given message is divided into numerous overlapping, redundant pieces, and these pieces are sent out over different routes, to different places, to be reunited before presentation to the receiver. Thus, even with the World Trade Center down – a major telecommunication processing center – e-mail was by and large able to get through. And in the days following the attacks, various e-mail lists and websites served as message boards for New Yorkers, who posted notes to let friends and the world know they were alive.

But the communications technology that performed the most heroic service on September 11 was the cellphone. Prior to that day, cellphones were under increasing criticism from a variety of sources. State and local governments were banning their use as handheld devices by drivers of automobiles. Passengers on planes were often forbidden to use them during flight. In a slightly different set of circumstances, cellphones might well have been prohibited from being carried on to planes altogether. Imagine how much worse September 11 would have been in that case – a fourth plane, hijacked by terrorists, crashing into the White House, or the Capitol Building, or perhaps second-hitting the Pentagon and destroying it, as had been done to the World Trade Center in New York City.

Instead, passengers on United Airlines Flight 93 were able to learn via their cellphones about the New York attacks, and take heroic action on their plane to prevent it from being used the same way. Who knows how many lives they saved, surely many.

And the events on United Airlines Flight 93 on that day were also of enormous symbolic importance. Todd Beamer, Jeremy

Glick, and the other people who stood up to the terrorists on that plane were the first to contest this new age of terrorism on behalf of civilization. At its very unleashing as a vehicle of mass destruction in America, international terrorism was told its victories could be punctured. History will record that the first action against terrorism in America on September 11 was not by the military, but by a handful of everyday people informed by the cellphone. Democracy's best weapon on that day was a pocket telephone that allowed passengers on a hijacked airline to find out the monstrous use of other hijacked planes, and move to prevent that on their doomed plane. If airplanes crashed deliberately into buildings was realspace at its worst, the heroism on United Airlines Flight 93 was the first ray of hope for realspace at its best.

The cellphone was also an instrument of personal heroic connection that day. Who among us, learning that we were likely to die unexpectedly, would not treasure a few minutes to say goodbye to those whom we loved most? The cellphone provided that gift to some passengers on two of the hijacked flights.

And the cellphone performed yet an additional service – by giving passengers the opportunity to talk one last time to loved ones, the cellphone gave the rest of the world some small inkling of what happened on those planes. In the days and weeks following September 11, recipients of those final calls appeared on television, and relayed to the world some of those conversations. Everyone who heard was in a sense on those planes, and witness to the anguish and the courage of the passengers. In the past, heinous mass murders were usually conducted with the eyes and ears of the world averted. On September 11, the cellphone shone a spotlight – or placed a microphone – into at least one of those hearts of darkness.

Thus the cellphone, which in normal times allows us to navigate the world, walk or ride through its precincts without losing touch with people or information, allowed us to navigate the world of terrorism on that day. The tragedy was that no medium – not the cellphone, not the Internet, no medium – was able to prevent it. This was in part because governments before September 11 did not know enough to employ media to prevent the attacks, and did not act to prevent terrorists from using their own cellphones, and the Internet, to plan the attacks.

And another medium – much older than cellphones and computers – was about to become the literal, physical delivery system of the next attack.

Anthrax and e-mail

The dissemination of anthrax through the U.S. Postal Service in the weeks following September 11 highlights another crucial difference between communication and transportation, and how the importance we place on one versus the other can shift in an age of terrorism.

In evaluating how much, in non-terrorist times, we should rely on communications versus transportation – on the Web and e-mail versus shopping malls and automobiles and letters in the mail – we could say that communication is preferable in cases where non-tangible information is conveyed. The Web, open twenty-four hours a day, seven days a week, in the worst imaginable weather, and from anywhere in the world, often beats the old-fashioned bricks-and-mortar library on this accounting. No one enjoys, or derives any benefit, from rushing to the library at 10:45 on a cold, rainy evening in February, only to find it closed at 10:30. On the other hand, the Web almost always comes in second to the in-person fruit-stand in purchase of tomatoes that are just the right ripeness. There is something in the heft, touch, smell of any fruit or vegetable that the Web cannot convey.

Even in better times, e-mail was preferable to street mail on most occasions. True, e-mail lacks the personal touch – it cannot be perfumed or kissed with lipstick – and one cannot include flower petals in an e-mail attachment. But e-mail, like every-thing else on the Web, is deliverable twenty-four/seven, and in circumstances far worse than rain or sleet. And it is deliverable, in days of terror, without danger of bodily infection.

In more whimsical times, I used to joke that online sex was safe sex. Condoms are not needed in cyberspace. I mentioned in Chapter 3 that virtual sex cannot make babies, and cannot spread venereal disease. These days, in the aftermath of September 11, we must add that e-mail cannot spread anthrax.

It is not giving in to terrorism to acknowledge that e-mail makes more sense than paper letters in envelopes in these trou-bled times. E-mail already made more sense, for reasons of convenience, in many situations. E-mail is instant, to anywhere in the world with an Internet connection, in one or a thousand copies. Now the reasons may also be more than ease and speed.

Terrorism wreaks its worst damage when it attacks the

physical – flesh-and-blood human beings, living and working and playing in real streets, real planes, real offices, real homes. Anything we can do to limit this real exposure without diminishing our lives seems advisable. If we can send and receive the same information digitally, virtually, electronically, as on paper, why choose physical paper?

This will not mean the end of the U.S. Postal Service, Federal Express, United Parcel Service, and the like. Physical objects cannot be sent online – only information – or words, images, and sounds. Everything else still requires sending the old-fashioned way, with the danger of terrorism that the ancient hand method now entails. But if we can reduce the number of items sent that way, and thereby limit terrorist entrée to items in post offices, mail sacks, mail trucks, and planes, does that not make sense?

Paper mail has been in decline, on the ropes, for decades anyway. Before the introduction of Express Mail (overnight delivery) several decades ago, the U.S. Post Office (as it was then called) had been losing money year after year. In 2001, even with Express Mail, and prior to September 11, the U.S. Postal Service was running a deficit. Chances are paper letters would have withered in favor of e-mail anyway in the next few decades. Why wait – for the likely inevitable – when we can send e-mail and reduce potential exposure to terrorist bacilli right now? E-mail is of course not immune to terrorist computer viruses which, if directed against medical, military, and similarly crucial activities, could result in loss of life. But even in such cases, the threat is once-removed – a virus disables a computer which disrupts a distribution of medication which results in death – in contrast to the directly deadly possibilities of an anthrax or other real bacillus or virus. In other words, the second-hand nature of communication (movement of information, in this case computer germs) in contrast to transportation (movement of physical objects, in this case real germs) is here a signal advantage.

The contention of this book has not been that physical inter-action and transport is always preferable to transmission of information, where we have a choice. Rather, I have argued in *Realspace* that we ought to be careful not to be so swept away by information technology and its wonders that we lose sight of unique and crucial advantages of physical presence in many human activities – here on Earth, and eventually in space. But the advent of e-mail made the advantages of physical delivery of paper mail negligible. The advent of anthrax in the mail makes them null and void.

Planes and rockets: reversal of symbols?

I did not see either of the planes hit the World Trade Center towers on September 11, directly with my own eyes. But like millions of people around the world, I saw those images again and again on television. And it changed the way I look at planes.

Having grown up in New York City with its major airports, I have been gazing at planes floating in the sky for years. Especially from my vantage point in the Bronx, I got good views of planes approaching La Guardia and Kennedy (once Idlewild) Airports from the north, languidly sailing down, so slowly as to almost look as if they were holding still, big butterflies or humming-birds suspended near the skyline, tasting the city, stopping in time. I found these freeze-frames to be comforting pieces of peace and grandeur, daily testaments to literally soaring human accomplishment, welcome breaks of a second or two from the hustle-bustle on the ground.

No more.

Now, when I see a plane hovering above, especially near a tall building, well . . . you know what I think. No humming-birds, butterflies, or soaring dreams. . . .

In addition to everything else, September 11 changed the symbolic tableau of skies.

Rockets used to evoke this response in many people, and perhaps still do. In the beginning and at several strategic places in *13 Days*, the 2000 movie about the Cuban missile crisis, we see clusters of rockets taking off. These were supposed to evoke the nightmare of Armageddon. But I have to confess that my very first feeling in seeing those scenes was joy – we're embarking to the stars!

Fortunately, intercontinental ballistic missiles have never been hurled across the world. London was a victim of terror by more local rocketry at the end of the Second World War – the dawn of rocketry as practical propulsion, and thus the dawn of our Space Age – and short-range missiles in ensuing decades have struck at the good and the bad in various parts of the world. And yet the transcendent act of terrorism which is defining at least the start of the twenty-first century was conducted by airplanes, not rockets – by flying machines descended from a technology a hundred years old – just as the atomic age and all of the fear it evoked was initiated by an airplane, not a rocket, over Hiroshima. The parallel between the two acts is technological and cultural – that is, the long-term cultural shadow of Hiroshima and the likely long-term impact of September 11 – not moral, since atomic weapons were used on Hiroshima and Nagasaki to defeat an enemy that had earlier attacked America. And perhaps the only reason why rockets were not used on those occasions was that they were not available (Germany, not America, had primitive rocket facilities at the end of the Second World War). But there does seem to be something about the plane, now clearer in the retrospect of September 11, that has made it a more ready weapon of mass destruction than the rocket, despite the rocket's generally worse press. Perhaps this is in some way because the airplane, in contrast to the rocket, was never conceived as a vehicle that could someday lift us not only off the earth but off the Earth.

The rocket's path to the spaceship, the starship, and the Universe, then, is still largely unobstructed by symbolic connection to mass calamity. Indeed, the rocket's cultural connection to disaster in the past half century has instead been mostly one of heroic loss of life, in Apollo 1, the Challenger, Columbia, and lesser known Soviet sacrifices. Human beings being what we are, that state of affairs may not last. It would be nice if we could reach some of the stars with our rockets before they are used for other ends down here on Earth.

Starport at the World Trade Center

Even more horrendous than the planes crashing into the WTC towers were the towers coming down. And for me, the worst part of those images was the tower with the antennae on top, slowly crumbling, crumpling back on itself like a tree in reverse to the Earth. To me, it seemed almost like a spaceship crumbling back on itself, as the U.S. Navy's Vanguard did in December 1957 when it tried to reach the sky, but fell back on its launchpad in flames. An awful replay – and far more destructive than the original.

One reason why the original was not so terrible, in addition to it causing no loss of life, is that the U.S.A. did successfully launch an artificial satellite – the Army's Explorer – a month later, in January 1958. This was only three months after the Soviet Union had jumpstarted the Space Age with the Sputnik. And so the race was on. And though it was between two superpower rivals, armed to the teeth with nuclear weapons and soon with intercontinental ballistic missiles, it had a very happy ending. There was never a nuclear exchange, not even a conventional war, between the two spacefaring nations. And before the Cold War was finally over, with the collapse of the Soviet Union at the end of the 1980s, these two superpowers between them had walked on the Moon, sent science craft to Venus and Mars, plied near-space in shuttles, and created the first long-lasting space station. Triumphs for humanity, each and every one of those.

It will take the world a lot more than a month to recover fully from the devastation of September 11. The rapid removal of the physical remains at Ground Zero is just the beginning.

What can we do to reverse fully the psychological, the symbolic devastation, once the physical has been repaired?

We can never recover the lives that were lost there. For that there is no repair, beyond never forgetting. But is there something we can do at the physical site, something more than rebuilding, something even more than memorializing, something that can in some sense exceed the original twin towers?

Various plans for reconstruction unveiled in the summer of 2002, nearly a year after September 11, were rightly given a cold reception. The plans were lackluster; nothing about them was memorable, or better than what used to be there.

We could build bigger towers, more than two towers, but that accomplishment would exceed the original only in the most literal, quantitative way.

What can we build there that would not only be physically immense, rise not only in steel but psychologically, symbolically? What can we build on the World Trade Center site that would someday rise to the sky not only symbolically, but actually, truly, really?

Let us look at this question from a different angle.

The antennae on top of the doomed WTC tower was of course an organ of communication. Should we build a bigger, better communication device of some sort there? Or can we perhaps improve upon communication itself, by constructing something on the World Trade Center site, on this now-hallowed ground, that will do more than move information?

Can we construct a center of transportation, a place that will move people – from this Earth to the stars?

Can we replace the falling tower and the antennae that looked like a rocket which crumbled, with a real starship that can actually take off for points beyond?

Certainly not in a year. Perhaps not even in a decade or a

century. Building a spaceport on the tip of Manhattan, at the World Trade Center site, were rockets powered by current chemical fuel or possibly nuclear fuel in the near future, would be much too dangerous. We cannot have rockets taking off on the thrust of explosions, however carefully managed, from Manhattan – not if we want New York City to continue as a center of population, business, and culture.

But what about a more distant future, when we have developed anti-gravity devices, or other non-explosive means, of lifting starships off this Earth?

At that point, whenever it comes, ships could rise from the World Trade Center site, from Ground Zero, to the Universe – to Ground Infinity – with no danger to the people near the launch site. And every time one of those starships rose, on its human mission to the Universe, it would be reversing the horror and terror of September 11. It would be lifting not only living people – the astronauts of the future – but the spirits of all those souls who perished, lifting them all to the stars.

Is that not at least one good answer we can give to those who sought to cripple our civilization?

And until such a launchpad can actually be made operational – and it could indeed be centuries before we develop anti-gravity lift-off technologies – we could construct at least the beginning of it, the foundation of it, the outline of it, at the World Trade Center site. It could take up just part of the site. It could be called the World Starport Center, its new structures pointing starward, straight up to the future.

A few last words about images, reality, and opportunity

Images were never thoroughly reliable reflections of reality. Certainly paintings were not. And even in the early decades of photography, photochemical concoctions abounded, purporting to show Mary Lincoln visited by her assassinated husband, and other ghostly events.

But the digital age has made photographic – and acoustic – manipulation almost as easy as typing on a computer. Whereas physical splicing was once required to rearrange sounds and moving images, and painstaking "cropping" to alter the still photograph, nowadays the conversion of sounds and images to a common binary code makes them almost as effortless to manipulate as counting to two.

Further, the human touch of the painting – its distance from literal reality, and its unreliability as a witness of that reality – was always obvious to everyone. In contrast, the photograph, videotape, and all modes of recording which originated prior to our digital age still seem, on first encounter, to be accurate mirrors of reality. These media carry their pre-digital imprimatur of authority, even though they are like painting and putty in the hands of any competent digital artist.

Of course, digital doctoring of photography, video, acoustic recording, and the like is presumably not yet able to erase all evidence of its deception. Experts can examine this or that tape of Osama bin Laden, and render a judgment as to its authenticity, about whether it has been tampered with or otherwise edited to alter its message. But sometimes, especially in courts of law, experts can disagree. When a photograph of O.J. Simpson was introduced at his civil trial (the one following his acquittal in the murder trial), showing him wearing a pair of shoes he alleged to no longer have had at the time of the photograph, his attorneys claimed the photograph had been altered – as indeed had been a photograph of O.J. Simpson in *Time*, to make him look, literally, more dark. (*Time* magazine claimed the darkening effect was unintentional.) The upshot is that, at a time when DNA is making tracking and identification of criminals and evidence more definitive, the older mode of photographic evidence has been digitally undermined, and brought into question.

Are we comfortable – if we agree that human movement into space is a worthy goal – to leave so much of those encounters to digital photography and relay of images? When we see that the image can be as much a concoction of the creator, as much an artifact of the processing device, as a reflection of what is really out there?

Now, more then ever, we need people in space. Our communications systems are already there. The events of September 11 show us how vulnerable we and our civilization are down here on Earth. We knew this already. We knew about Chinese rockets and printing, Alexandrian moving images, Aztec wheels, and many other devices that went little distance or nowhere at the times of their invention.

Perhaps, because the democracies of the West had been so victorious through most of the twentieth century, we believed that our civilization, unlike the Aztecs and the Alexandrians and so many others, would last forever. We thought, perhaps, that we had all the time in the world to go to space.

Now, perhaps, we know a little different. Western civilization in the aftermath of early twenty-first-century terrorism may even emerge stronger. But we also know we are not invulnerable.

So let us use our strength, our awareness of our mortality as a civilization, to do something truly lasting and Earth-shaking for humanity. Let us join with the peoples and cultures of this planet, the diversity of its perspectives and religions and science, so we can better leave it – not behind, but as a springboard to something greater.

It's time to go from our screens to the stars.

SELECT BIBLIOGRAPHY

with annotations

13 Days (2000) Movie directed by Roger Donaldson about the Cuban missile crisis.

> Imaginary scenes of rockets taking off are supposed to evoke a sense of Armageddon, narrowly averted in the crisis. But my initial reaction, when I saw the rockets, was a sense of joy, because they made me think of humanity reaching out to the stars. In the twentieth and the twenty-first century thus far, airplanes, not rockets, have been the most frequent conveyors of mass destruction. (See Chapter 11 in *Realspace* for more on the symbolism of planes and rockets.)

2001 (1968) Movie directed by Stanley Kubrick, written by Kubrick and Arthur C. Clarke, based on Clarke's story "The Sentinel," first published in 1951.

A significant portrayal of humanity in near space, with cosmic implications. Clarke wrote many stories like this in the 1950s (see Clarke, *Childhood's End*, below); by the late 1960s, they had become more frequent on television and in films than in literature.

Agassi, Joseph (1968) *The Continuing Revolution*, New York: McGraw-Hill.

Contains a reprint of the Johannes Stradanus (1523–1605) engraving of a bustling sixteenth-century print shop; the original caption advises, "Just as one voice can be heard by a multitude of ears, so single writings cover a thousand sheets." The analogy is a little awkward – "so single writings can be seen by a thousand eyes" would have been better balanced – but it is nonetheless an apt description of the way the printing press humanized writing, by making it more like speech. Cyberspace continued this process, by giving writing the same speed of transmission as speech in a face-to-face conversation. But when cyberspace actually takes the place of in-person conversation, is this making the world more or less human? (In addition to containing the Stradanus engraving, Agassi's book also offers a good account of the growth of knowledge through trial and error.)

Appleton, Victor (1914) *Tom Swift and his Photo Telephone*, New York: Grosset & Dunlap.

Videophone makes an appearance in science fiction less than four decades after the invention of the telephone in 1876; its imminent replacement of the telephone was predicted periodically throughout the twentieth century. The fact that this has still not happened demonstrates the need for collateral technologies and conducive social attitudes in the adoption of inventions. (See also Burlingame, below.)

Asimov, Isaac (1951–1953) *Foundation; Foundation and Empire; Second Foundation*, New York: Gnome.

A series of three science fiction novels, comprised mostly of shorter works published in the 1940s, that won the World Science Fiction Society's "Hugo" Award for the best trilogy of "all time" (competition for that award included Tolkien's *Lord of the Rings*). These novels represent one of the epitomes of future humanity comfortable in worlds beyond our planet: the fate of a human empire spanning the galaxy hangs in the balance, as protagonists travel from one star system to another with the ease that we now travel from city to city on Earth. (See also additions to the *Foundation* series, written by Asimov in the 1980s and 1990s.)

Asimov, Isaac (1953, 1956, 1983) *The Caves of Steel; The Naked Sun; The Robots of Dawn*, New York: Doubleday.

In this trilogy of science fiction, Asimov develops his "three laws of robotics," first introduced in short stories in the 1940s: (1) a robot can never harm a human or by inaction allow harm to happen, (2) a robot must follow all human commands, except when they contradict the first law, (3) a robot must always act to insure its own survival, except when such actions contradict the first two laws. Thus, a robot would be compelled by its programming to follow a human command to jump off a roof – as long as there were no humans in danger of being hit by the falling robot. These laws were designed to show how robots could be used safely, with no danger to humans, on Earth and in space. As a writer of riveting fiction, Asimov of course spun stories that did not quite work out that way. Moreover, in considering robots as our emissaries to the stars, I have come upon the following ethical quandary: the more intelligent our robots become the better they can serve us, but the closer their attainment of human sentience, the less ethical right we have to use them as slaves. (See also Asimov's subsequent robot novels in the 1990s, where he introduces a "zeroth law": the good of the human species as a whole outweighs the robotic need to protect the life of any individual.)

Barlow, John Perry (1995) "Is There a There in Cyberspace?" *UTNE Reader*, March.

Contrasts cyberspace to "the physical world which I call 'Meatspace'" – or what I call "realspace" in this book. Barlow observes accurately "the fundamental and profound difference between information and experience," and the lack of the latter, in diversity and sensory forms, in cyberspace. But I prefer "realspace" for this full-contact world, because (1) there is more in this world than just flesh and blood, and (2) the word "real" conveys correctly a superiority to cyberspace. (For those who enjoy philosophy, we might say that realspace conveys the ontological priority of the physical world; in plainer English, it gets at what we cannot live without.) (See also Elmer-Dewitt, below, for more on Barlow and cyberspace.)

Bazin, André (1967) *What Is Cinema?* trans. H. Gray, Berkeley, CA: University of California Press.

First published as *Que-est-ce que le Cinema?* in four volumes, 1958 to 1965, Bazin here defines the photograph as rescuing an image from "its proper corruption" in time. See Chapter 2 of *Realspace* for a discussion of how, once pulled out of time, the image on the photograph is reinserted into the world of motion by the pockets, handbags, and various carrying contrivances of people moving through the physical world. In the age of photography, communication and transportation were still very much in balance. (The advent of motion photography would soon restore life to the frozen image in the world of information.)

Burlingame, Roger (1959) "The Hardware of Culture," *Technology and Culture*, 1, 1: 11–28.

A classic, cogent argument about the need of inventions for "collateral technologies" in order to succeed; Leonardo's were not actualized in his time because of this lack. What collateral technologies – and social inducements – are missing in our attempts to build a spacefaring civilization?

Campbell, Donald T. (1974) "Evolutionary Epistemology," in P. Schilpp (ed.) *The Philosophy of Karl Popper*, La Salle, IL: Open Court.

Campbell here details perception in organisms and cognition in humans as providing increasingly vicarious representations of reality – the increasing distance from reality has the dual effect of being safer (full physical contact – the way the amoeba knows its world – is more dangerous than seeing) and more prone to error (tactile illusions are far less common than optical illusions). The increase in safety drives the evolution of vicarious modes, culminating in cognition in humans: it is safer to think about something dangerous than to see or hear it. I pick up Campbell's model, and apply it to technology (see my *Mind at Large*, and *The Soft Edge*, both below): looking at Mars through a telescope is certainly safer than traveling there, and it gives us much more information than may be obtained by just thinking about Mars. But this knowledge is still less than the knowledge to be gained by literally walking on Mars, and therein resides the need for realspace in outer space. (As for cyberspace on Earth, it is undeniably safer than any mode of physical transport. But what is lost when we travel via electrons rather than cars, trains, boats – and feet?)

Carson, Rachel (1962) *Silent Spring*, Boston, MA: Houghton Mifflin.

Marshall McLuhan suggested that Sputnik (1957), the first sustained technological excursion off this planet, ushered in the age of ecology, since the extraterrestrial view gave us a sense of the precariousness of our planet in the void. Carson's book is usually cited as the initiating manifesto of current concern for ecology, so the timing is right. Focus on ecology and identification of technology as a despoiler was also one of the factors in the 1960s and 1970s that soured much of public opinion on all things technological, including ships to lift us off this planet. (See McLuhan and Bruce Powers, *The Global Village* (New York: Oxford University Press, 1989) for more on Sputnik and the Earth as "an object of art.")

Clarke, Arthur (1953) *Childhood's End*, New York: Harcourt.

A classic science fiction novel about humanity's destiny in the stars. On the balance of transportation and communication, note that Clarke is also the author of "Extra-Terrestrial Relays" (*Wireless World*, October 1945, pp. 305–308), where he proposes – in an essay, not fiction – that artificial satellites in geostationary orbits can be used for global communication. The first transatlantic television signal was transmitted by the Telstar satellite in 1961.

Ei, Rokusuke and Nakamura, Hachidai (1961) "Ue O Muite Aruko."

This Japanese song was recorded by Kyo Sakamoto, and became a hit record in the U.S.A. – under the title "Sukiyaki," which had no connection to the lyrics, other than that they were in Japanese. The real title translates as "I Look Up When I Walk," and the first line contains the punchline, "So The Tears Won't Fall." This song could have been an anthem for the inspiration and travails of reaching for the stars, but I cite it in Chapter 2 of *Realspace* as an example of how easily communication can be taken out of context and mislead: millions of Americans thought (and continue to think) that this song is about cuisine. Cyberspace not grounded in realspace can be seen as a world of communication out of context.

Ellul, Jacques (1964) *The Technological Society*, trans. J. Wilkinson, New York: Vintage.

First published in 1954 as *La Technique*, this classic, vivid, unrelenting critique sees the human being stuck "like a fly on the flypaper" of technology. Ellul's attack is comprehensive – all technology, throughout history, is indicted – and thus self-defeating. (How can we take an attack on all technologies, including books, seriously, when presented in a book?) Ellul also gives short shrift to the profound life-enhancing effects of many technologies, including those of transport and communication. But his "fly on the flypaper" is uncomfortably reminiscent of human eyes glued to computer screens – relievable any time by a walk in realspace, if we choose. (The cellphone helps, by allowing us to be in contact with at least part of the world of information as we walk through the world around us. But it also puts us constantly on call, which is another kind of glue.)

Elmer-Dewitt, Philip (1995) "Welcome to Cyberspace," *Time*, special issue, spring.

Quotes John Perry Barlow's definition of cyberspace as the "place where people are when they are on the telephone." For earlier renditions of essentially the same apt view, see Marshall McLuhan's observation that we become "discarnate" when watching television, listening to radio, or talking on the phone ("A Last Look at the Tube," *New York Magazine*, April 3, 1978, p. 45), and Edmund Carpenter's note that "electricity has made angels of us all ... spirit freed from flesh, capable of instant transportation anywhere" (*Oh, What a Blow that Phantom Gave Me!* New York: Bantam, 1972/1973, p. 3). Significantly, although the telephone offered the same discarnate, "angelic" environment as cyberspace for a hundred years, no one thought of "phonespace" as a place to inhabit or visit for business or pleasure. This was because the telephone, unlike cyberspace, was seen as an adjunct or an extension, not a replacement, of realspace. See my *Digital McLuhan*, below, for more on McLuhan and Carpenter and discarnate/angelic denizens of electronic realms; see my *The Soft Edge*, below, for discussion of the telephone as an essential precondition of cyberspace; see Barlow, above, for more of his views on cyberspace.

Freud, Sigmund (1930) *Civilization and Its Discontents*, trans. J. Riviere, New York: Cape and Smith.

Freud here defines writing as the "the voice of an absent person" – as good a definition as any for all communication in which the sender of information is physically separated from the receiver. Writing in effect was thus the beginning of cyberspace, and wrought miracles in a world which was too bound by realspace. Printing, telegraphy, photography, the phonograph, motion photography, radio, television, and computers through the mid-1990s continued to liberate communication from the confinements of realspace. But has the shift gone too far? Are we now more in need of voices of people present? (See my *The Soft Edge*, below, for more on the liberating evolution of media. See also Freud, "The Relation of the Poet to Daydreaming," in B. Nelson, ed., *On Creativity and the Unconscious* (New York: Harper & Row, 1958), and Freud, *Totem and Taboo* (New York: Vintage, 1918), for more on his equation of the creative processes of the child, artist, neurotic, and pre-technological primitive discussed in Chapter 8 of *Realspace*.)

Gibson, William (1984) *Neuromancer*, New York: Ace.

Acclaimed science fiction novel, one of the first in the "cyber-punk" genre. By the 1980s, the cutting edge of science fiction had turned its back on space and the sense of wonder about life beyond this planet that had animated science fiction from the 1930s through the 1950s (see e.g. Asimov, above), and moved instead to dystopic visions of the Information Age. *Neuromancer* also contains the first printed mention of "cyberspace."

Heinlein, Robert (1956) *Time for the Stars*, New York: Charles Scribner's.

Heinlein is probably best known for his military science fiction – such as *Starship Troopers* (1959), made into a movie by Paul Verhoeven in 1997 – but *Time for the Stars* is a classic in the human-ization of space, more specifically the impact of space travel on family life. One twin travels to the stars at close to the speed of light; the other stays on Earth. Due to relativity effects (time is supposed to slow as speed approaches that of light), the traveler ages four years while his Earth-bound twin ages seventy, and they keep in intermittent contact throughout. Marketed (and marginalized) as a "juvenile" or "young adult" novel – likely because the twins are age 17 at the start of the story – *Time for the Stars* is more adult than any fiction that takes place solely on Earth, in its exploration of family relations strained by cosmic relations of space and time. But, then again, as I argue at the end of Chapter 8 (this volume), children are often more adult than many adults in their conceptions of outer space. (For a shorter, exquisite exposition of a family split by star travel and time distortion, I recommend Paula May's "Dilating the Paradox," *Analog: Science Fiction and Fact*, mid-December, 1994.)

Levinson, Paul (1979) "Human Replay: A Theory of the Evolution of Media," Ph.D. diss., New York University.

I develop my "anthropotropic" theory of media evolution in this work – as media evolve, they become more human, less artificial, in their performance ("anthropo" = human; "tropic" = towards). Thus, color photography and television emerged because we see the world in color, not black-and-white. Cyberspace (not discussed by me in 1979) is an anthropotropic improvement over paper mail and mail-order shopping; but is it a step backwards, a diminishment of humanity, when it replaces face-to-face communication in the real, off-line world?

Levinson, Paul (1988) *Mind at Large: Knowing in the Technological Age*, Greenwich, CT: JAI Press.

Defines and explains human technology as an embodiment of ideas.

Levinson, Paul (1992) "An Easter Theory of Technology," in *Electronic Chronicles: Columns of the Changes in Our Time*, San Francisco, CA: Anamnesis Press.

> Wonders if the emergence of modern technological civilization in Roman Catholic Western Europe was encouraged by the unusual Roman Catholic view that Christ returned not only in spirit but in flesh, thus setting the stage for an appreciation of physical miracles via technology. (See Pagels, below, for an explanation of why the Roman Catholic interpretation was so unusual.)

Levinson, Paul (1997) *The Soft Edge: A Natural History and Future of the Information Revolution*, London and New York: Routledge.

> Survey and analysis of the communications side of the communications/transportation partnership throughout history, currently culminating in cyberspace. *Realspace* examines the ways that cyberspace can unbalance the partnership in favor of communications, and urges remedies.

Levinson, Paul (1999) *Digital McLuhan: A Guide to the Information Millennium*, London and New York: Routledge.

> The digital age is pertinent to the work of Marshall McLuhan (1911–1980) in two ways: his ideas have become more comprehensible, and they provide the best available framework for making sense of life online. They were under-appreciated, misunderstood, and misapplied in the 1960s because he was in effect writing about a logical development in communications, cyberspace, which had not yet come into being. He wrote little about "realspace," because, prior to cyberspace, no one missed it. (But his "tetrad" assessment of media, discussed in *Digital McLuhan*,

made clear that the development of new media often brings to our attention older systems recently taken for granted. For more on the evolution of media, see my "Human Replay" and *The Soft Edge*, above.)

McLuhan, Marshall (1962) *The Gutenberg Galaxy*, New York: Mentor.

The first mass mention of the "global village" (earlier coined by McLuhan in his "Report on Understanding Media," typescript published by the National Association of Educational Broadcasters, US Department of Health, Education, and Welfare, Washington, DC, June 30, 1960). The global village in McLuhan's time was not quite global (although Telstar delivered the first transatlantic television signal in 1961, widespread use of satellites for conveyance of television around the world was at least two decades away), and not a village whose denizens communicated easily among themselves (television is a one-way medium). The fully global, interactive village awaited the arrival of cyberspace. Inasmuch as transportation will always lag behind communication in speed – at least until we can move people as well as electrons at the speed of light – the global village in cyberspace will always have an advantage over global communities in realspace. In this case, then, cyberspace is not encroaching upon realspace, but providing an environment unattainable by transportation.

Minsky, Marvin (1986) *The Society of Mind*, New York: Simon & Schuster.

Minsky's view that our minds are just "meat machines" – by which he means that there is nothing non-physical about minds – is often quoted. Here, he discusses "brain machines," by which he intends the same point. In either rendition, my response is: if brains are just meat or machines, then they are like no other meat or machines known to science (except insofar as human brains resemble brains of gorillas and chimps – yet even in those brains, we know of no instances in which they have given rise to works of Shakespeare or even the worst human writer in the world). This difference does not prove that our brains are animated by spirits or souls; but it does speak to the profound difference between our brains and meat machines. The relevance of all of this to *Realspace* is that we need human brains or their full equivalent out in space; true machines – the current state of robots – won't do, if only because they will miss the poetry of the universe.

Oboler, Arch (1956) *Night of the Auk*.

After the great auk, an extinct bird, this play ran for eight nights on Broadway (with a televised version, starring William Shatner, in 1960). Earth's first crew to the Moon receives a devastating surprise on their way home: nuclear war has broken out on the planet, and their docking station has been destroyed. They, along with the Earth's hopes for the future, are stranded, sentenced to die, at the very moment of our greatest success. In reality, Sputnik was launched a year later; Neil Armstrong and Buzz Aldrin made it to the Moon and back in 1969; and so far we have managed to avoid nuclear war. But have our efforts in space gone much further than the auk's?

Pagels, Elaine (1979) *The Gnostic Gospels*, New York: Random House.

Argues that, unlike unsuccessful, rival gnostic views of the Resurrection which claimed that Christ returned only in spirit, the prevailing Roman Catholic interpretation was distinguished by its insistence that Christ also returned in the flesh. (See Levinson (1992) above, for the possible relevance of this to the rise of technology in the West.)

Plato (n.d.) "Phaedrus", in B. Jowett (trans.) *The Dialogues of Plato*, New York: Scribner, Armstrong.

Socrates here attacks the written word as a "misbegotten" image of the spoken original, unamenable to live dialog. Cyberspace has given the written word this missing interactivity, but does this now make it preferable in all cases to conversation with complete, human presence?

Popper, Karl (1945) *The Open Society and its Enemies*, London: George Routledge & Sons.

Popper deftly shows how Plato's philosopher-king – dictatorial government by the wisest – leads to Nazi Germany and the Soviet Union. The relevance of this, in 2003, to space: although the totalitarian societies of the twentieth century were crucial in initial departures from this planet, and although democracy in the U.S.A. has been an unreliable supporter of efforts in space, we are still better advised to work with inefficient democracy. Philosopher-kings are equally efficient in doing things good and evil.

Roland, Alex (1994) "How We Won the Moon," *New York Times Book Review*, July 17: 1, 25.

One of the best-reasoned arguments for why robots can suffice as our emissaries to the planets and stars – that is, why humans are not necessary in our excursions beyond this planet. See Chapter 10 of *Realspace* for why this is wrong.

Sabrina (1954)

One of the very first movies to feature a cellphone (used by the character played by Humphrey Bogart, early in the film, in the back of his limousine). See Chapter 4 of *Realspace* for the cellphone as an early expression of the need for people to get up from their chairs and screens and out into the world.

Star Trek (1966–1969) Created by Gene Roddenberry.

The original TV series – conceived by Roddenberry as a "Wagon Train to the Stars" (*Wagon Train* was a highly successful Western series on TV from 1957 to 1965) – which engendered (as of 2003) four subsequent TV series and ten movies. *Star Trek* and the galactic civilizations it portrayed came at an opportune time for space enthusiasts; by the mid-1960s the sense of wonder in science fiction books and stories had waned, at least insofar as space. *Star Trek* (and *Star Wars*) helped keep the flame of popular interest in space alive, but lit no new fires.

Star Wars (1977) Directed by George Lucas.

The first *Star Wars* movie, Episode 4 in the saga, later renamed *A New Hope*, with four sequels as of 2003 (actually, two sequels and two "prequels"). Had much the same positive impact on space in popular culture as *Star Trek*, with the same limitations.

Thompson, William Irvin (1973) *Passages About Earth: An Exploration of the New Planetary Culture*, New York: Harper & Row.

A literally symbolic argument against the space race, or in any event rushing further or more frequently into space: Thompson contends that, having reached the Moon, we have changed our self-conception and culture for the better, and thus have no need to get back there or any place else in space anytime soon. The part about reaching the Moon having changed us for the better is true, but the rest of the argument fails, for at least three reasons: (1) We do not live by culture alone; we need resources beyond this planet. (2) Culture is better served by continuing excursions in reality than a diet of replays. (3) The rise and fall of civilizations in the past, and missed technological opportunities in cultures that have survived, suggest that the present is the best time to pursue our goals in space. All three factors – tangible dividends in space, culture nourished by reality, taking advantage of the present at hand – are also arguments for realspace over cyberspace.

Time magazine (1994) June 27.

The infamous issue with the darkened picture of O.J. Simpson on the cover. Photography was never beyond alteration – cropping, superimposition of images not in the original scene, have been staples almost since its inception. But until the advent of digital media, the camera provided a much more reliable image of reality than painting, which captured not the world as it was but as the artist chose to render it. The situation is now reversed: pixels are easier to manipulate than pigments, and the result is usually much more convincing, since the components of the digital image are reflections of reality. This almost effortless, persuasive revisability of information in cyberspace is one of its prime differences from realspace.

Watt-Evans, Lawrence (ed.) (1991) *Newer York*, New York: ROC.

An anthology of then-new science fiction stories depicting the future of New York City. The title appeals to me as an example of how a city in space, or on another planet, could be named in a way that would carry the connotations of home (in the way New York did this, in the first place, *vis-à-vis* York). New Chicago and Newer London would also work.

INDEX

INDEX

INDEX

INDEX

INDEX

Vanguard (failed U.S. space
    launch) 147
Vatican II Council 99
video conferencing 44
video-cassette recorder (VCR) 48
videophone 53–5
Vietnam War 88
virtual sex 40, 142

war 11, 87–9, 97, 115–16
    nuclear war 69–70, 88
    Second World War 27, 30, 86,
        87, 145–6
    see also Cold War

Watt-Evans, Lawrence 114, 169
Western Union (Telegraph) 53
window shade 48
Winner, Langdon 48
wish-fulfillment 105–8
World Trade Center (WTC) 136,
        139, 144, 146–8
    reconstruction as space port
        148–9
World War II see Second World
    War
Wozniak, Steve 87
Wright brothers 87
writing 19–21
Wynn, Marie 34

Printed and bound by CPI Group (UK) Ltd, Croydon, CR0 4YY

01/11/2024

01782630-0001